VOLUME IV
Field Effect Devices
Second Edition

MODULAR SERIES
ON SOLID STATE DEVICES

Robert F. Pierret and Gerold W. Neudeck, Editors

VOLUME IV
Field Effect Devices

Second Edition

ROBERT F. PIERRET

Purdue University

ADDISON-WESLEY PUBLISHING COMPANY

READING, MASSACHUSETTS
MENLO PARK, CALIFORNIA · NEW YORK
DON MILLS, ONTARIO · WOKINGHAM, ENGLAND
AMSTERDAM · BONN · SYDNEY · SINGAPORE
TOKYO · MADRID · SAN JUAN

This book is in the
Addison-Wesley Modular Series on Solid State Devices

Library of Congress Cataloging-in-Publication Data

Pierret, Robert F.
　　Field effect devices / by Robert F. Pierret–2nd ed.
　　　　p.　cm. — (Modular series on solid state devices)
　　Includes bibliographical references.
　　ISBN 0-201-12298-7
　　1. Field-effect transistors.　I. Title.　II. Series
TK7871.95.P53 1990　　　　　　　　　　　　　89-49075
621.381'5284 — dc20　　　　　　　　　　　　　CIP

Reprinted with corrections October, 1990

Foreword

Solid state devices have attained a level of sophistication and economic importance far beyond the highest expectations of their inventors. By continually offering better performing devices at lower cost per unit, the electronics industry has penetrated markets never before addressed. A necessary prerequisite to sustaining this growth initiative is an enhanced understanding of the internal workings of solid state devices by modern electronic circuit and systems designers. This is essential because system, circuit, and IC layout design procedures are being merged into a single function. Considering the present and projected needs, we have established this series of books (the Modular Series) to provide a strong intuitive and analytical foundation for dealing with solid state devices.

Volumes I through IV of the Modular Series are written for junior, senior, or possibly first-year graduate students who have had at least an introductory exposure to electric field theory. Emphasis is placed on developing a fundamental understanding of the internal workings of the most basic solid state device structures. With some deletions, the material in the first four volumes is used in a one-semester, three credit-hour, junior-senior course in electrical engineering at Purdue University. The material in each volume is specifically designed to be presented in 10 to 12 fifty-minute lectures.

The volumes of the series are relatively independent of each other, with certain necessary formulas repeated and referenced between volumes. This flexibility enables one to use the volumes sequentially or in selected parts, either as the text for a complete course or as supplemental material. It is also hoped that students, practicing engineers, and scientists will find the series useful for individual instruction, whether it be for reference, review, or home study.

A number of the standard texts on devices have been written like encyclopedias, packed with information, but with little thought as to how the student learns or reasons. Texts that are encyclopedic in nature are often difficult for students to read and may even present barriers to understanding. By breaking the material into smaller units of information, and by writing *for students,* we have hopefully constructed volumes which are truly readable and comprehensible. We have also sought to strike a healthy balance between the presentation of basic concepts and practical information.

Problems which are included at the end of each chapter constitute an important component of the learning program. Some of the problems extend the theory presented in the text or are designed to reinforce topics of prime importance. Others are numerical problems which provide the reader with an intuitive feel for the typical size of key parameters. When approximations are stated or assumed, the student will then have confidence that cited quantities are indeed orders of magnitude smaller than others. The end-of-chapter problems range in difficulty from very simple to quite challenging. In the second edition we have added a new feature — worked problems or *exercises*. The exercises are collected in Appendix A and are referenced from the appropriate point within the text. The exercises are similar in nature to the end-of-chapter problems. Finally, Appendix E contains volume-review problem sets and answers. These sets contain short answer, testlike, questions which could serve as a review or as a means of self-evaluation.

Reiterating, the emphasis in the first four volumes is on developing a keen understanding of the internal workings of the most basic solid state device structures. However, it is our hope that the volumes will also help (and perhaps motivate) the reader to extend his knowledge — to learn about the many more devices already in use and to even seek information about those presently in research laboratories.

> Prof. Gerold W. Neudeck
> Prof. Robert F. Pierret
> Purdue University
> School of Electrical Engineering
> W. Lafayette, IN 47907

VOLUME IV — SECOND EDITION

Everything is the same, nothing is the same. The foregoing perhaps best describes the second edition of Volume IV. Although very little has been deleted from the first edition, the volume has nevertheless undergone a massive revision. In addition to the necessary maintenance and updating, the revision reflects the experience of the author that abbreviated, and at times, reduced-level coverage of the material may be desirable.

There is now a General Introduction at the beginning of the volume and a new chapter treating modern FET structures at the end of the volume. In between there has been a significant rearrangement of the subject matter. First-edition Chapters 2 and 3, respectively devoted to ideal MOS device statics and $C-V$ characteristics, have been combined into a single simplified chapter entitled "MOS Fundamentals." The more complex portions of the first-edition chapters — the exact electrostatics and the corresponding $C-V$ analysis — have been placed into second-edition Appendixes B and C. Also, the major portion of the original MOSFET chapter has been repositioned *before* the discussion of nonidealities. Combined with modifications to assure Chapter 1 (J-FET and MESFET) and Chapter 4 (Nonideal MOS) can be skipped without loss of continuity, the restructuring establishes a number of viable options in the presentation of the subject matter. Specifically, the General Introduction plus Chapters 2, 3, and 5 provide a solid, coherent introduction to field-effect devices. An instructor could even omit Chapter 5 if minimal coverage is desired. The inclusion of Chapter 4, on the other hand, decidedly increases the level and breadth of the presentation. Further sophistication can be achieved by incorporating the material from Appendixes B–D. Naturally, the most pedagogically sound coverage includes Chapter 1.

In addition to restructuring of the volume, there are a number of other noteworthy changes and enhancements. As mentioned in the general Foreward, both an appendix of exercises and an appendix containing volume-review problem sets plus answers have been added to the volume. (Unlike Volumes I–III, the first edition of Volume IV did not contain volume-review problem sets.) Conversely, acknowledging difficulties associated with their use, certain symbols prominent in the first edition have either been deleted or are used sparingly in the second edition. The normalized potentials, (U, U_F, U_S) are found only in the second-edition appendixes, while V'_G (the voltage applied to the gate of an ideal device) now appears only in Chapter 4. The relation $x'_o \equiv (K_S/K_O)x_o$, which invariably led to errors in numerical computations, has been totally eliminated. Also eliminated is the use of $T = 23°C$ for room temperature. All computations and examples in the second edition employ the standard $T = 300$ K. The volume, of course, contains new and revised homework problems. In the author's opinion, the second edition of Volume IV is truly both new and improved.

—R. F. Pierret

Contents

General Introduction

Historically, the field-effect phenomenon was the basis for the first type of solid-state transistor ever proposed. Field-effect transistors predate the bipolar junction transistor by approximately 20 years. As recorded in a series of patents filed in the 1920s and 1930s, J. E. Lilienfeld in the United States and O. Heil working in Germany independently conceived a transistor structure of the form shown in Fig. 1. The device worked on the principle that a voltage applied to the metallic plate modulated the conductance of the underlying semiconductor, which in turn modulated the current flowing between ohmic contacts A and B. This phenomenon, where the conductivity of a semiconductor is modulated by an electric field applied normal to the surface of the semiconductor, has been named the *field effect*.

The early field-effect transistor proposals were of course somewhat ahead of their time. Modern-day semiconductor materials were just not available and technological immaturity, in general, retarded the development of field-effect structures for many years. A practical implementation had to await the successful development of other

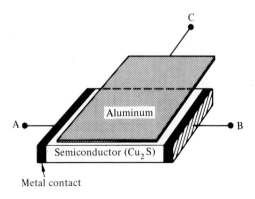

Fig. 1   Idealization of the Lilienfeld transistor.

solid-state devices, notably the bipolar junction transistor (BJT), in the late 1940s and early 1950s. The first modern-day field-effect device, the junction field effect transistor (J-FET), was proposed and analyzed by W. Shockley in 1952 [1]. Operational J-FETs were subsequently built by Dacey and Ross in 1953 [2]. As pictured schematically in Fig. 2 (which was adapted from an early Dacey and Ross publication [3]), *pn* junctions replaced the metallic plate in the Lilienfeld structure, the A and B contacts became known as the source and drain, and the field-effect electrode was named the gate. In the J-FET it is the depletion regions associated with the *pn* junctions that directly modulate the semiconductor conductivity between the source and drain contacts.

A key component of present-day microelectronics, the metal-oxide-semiconductor field-effect transistor (MOSFET), achieved practical status during the early 1960s. The now-familiar planar version of the structure pictured in Fig. 3, a structure with thermally grown SiO_2 functioning as the gate insulator, a surface-inversion channel, and islands doped opposite to the substrate acting as the source and drain, was first reported by Kahng and Atalla in 1960 [4]. Following intense development efforts, commercial MOSFETs produced by Fairchild Semiconductor and by RCA became available in late 1964. Figure 3 originally appeared on the cover page of the 1964 applications bulletin describing the early Fairchild MOSFET [5].

A major development occurring in the latter half of the 1960s was the invention by Dennard [6] of the one-transistor dynamic memory cell used in the random access memory (DRAM). The DRAM cell (see Fig. 4) is an integrated combination of a

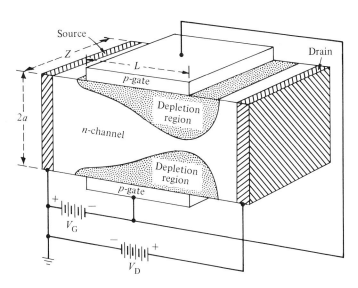

Fig. 2   Schematic diagram of the junction field effect transistor (G. C. Dacey and I. M. Ross, "The Field Effect Transistor," *Bell System Technical Journal*, Vol. 34, No. 6, Part II, 1955, pp. 1149–1189. Copyright © AT&T. Reprinted by special permission.)

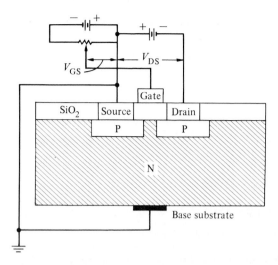

Fig. 3   Cross section of a MOSFET with correct biasing polarity (From *Fairchild Semiconductor Application Bulletin,* November 1964, APP 109, "Applications of the Silicon Planar II MOSFET," by John S. MacDougall. Reprinted by permission of National Semiconductor.)

charge storage capacitor and a MOSFET utilized as a switch. A quote from a recent review article [8] best underscores the significance of the invention and its impact. "The 1-T DRAM cell is now the most abundant man-made object on this planet earth."

The early 1970s saw the introduction of still another significant field effect structure — the charge-coupled device (CCD). Physically, a CCD might be viewed as a MOSFET

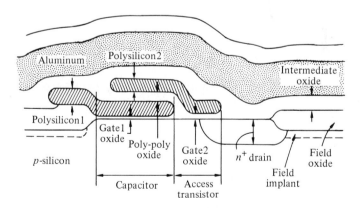

Fig. 4   Schematic cross section of a modern DRAM cell. (From Wyns and Anderson, [7], © 1989 IEEE.)

with a segmented gate. As pictured in Fig. 5, the proper application of biases to the CCD gates induces the systematic movement or transfer of stored charge along the surface channel and into the device output. The image sensing element in the camcorder, a combination home TV camera and video recorder, is a CCD containing a two-dimensional array of gates. Stored charge is produced in proportion to the optical image incident on the two-dimensional gate array. The electrical analog of the optical image is subsequently transferred to the output and converted into an electrical signal.

The invention, development, and evolution of field-effect devices continues to the present. Notably, in progressing through the 1980s, complementary MOS or CMOS

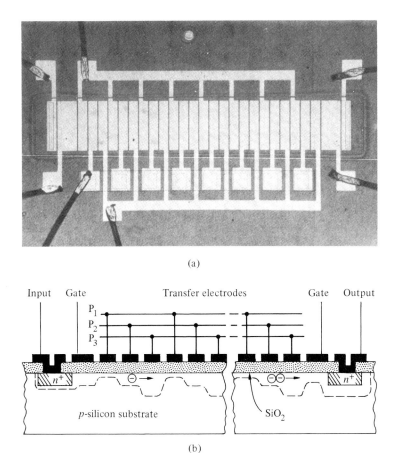

(a)

(b)

Fig. 5 First charge-coupled device comprising eight three-phase elements and input-output gates and diodes, shown (a) in plan view and (b) schematically in its cross-sectional view. (From Tompsett et al., [9].)

became the integrated circuit technology of choice. Figure 6 shows a cut-away view of a fundamental building block in CMOS circuitry, the CMOS inverter. The term "complementary" denotes the fact that, unlike other MOS circuit technologies, both n-channel (electron current carrying) MOSFETs and p-channel (hole current carrying) MOSFETs are fabricated on the same chip. Although CMOS came to the forefront in the 1980s because of its lower power dissipation and other circuit-related advantages, the change was evolutionary — CMOS is not a new circuit technology. It was initially conceived by Wanlass [10] in the early 1960s and was even mass produced as part of LED watch circuitry in 1972.

Whereas the use of CMOS exemplifies continuing evolutionary changes, there is also continuing field-effect device invention and development. The modulation doped field effect transistor (MODFET) pictured in Fig. 7 is an excellent example. Although the electron transport properties of GaAs are superior to silicon, GaAs technology lacks a quality homomorphic (Ga or As containing) insulator comparable to SiO_2. Moreover, the deposition of other insulators on GaAs has consistently failed to produce devices with acceptable electrical characteristics. The emergence of sophisticated deposition techniques such as molecular beam epitaxy (MBE), however, has made it possible to grow layers of different lattice-matched semiconductors on top of each other. The alloy semiconductor AlGaAs appearing in Fig. 7 has a larger band gap than GaAs and essentially assumes the role of "insulator" in the MODFET structure. The MODFET promises to be a strong contender for use in the high-speed logic circuits of the 1990s.

This volume is intended to be an introduction to field-effect devices. The more basic members of the field-effect-device family are examined in some detail to expose the

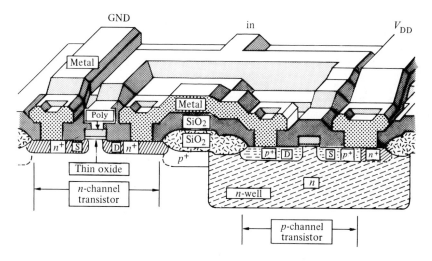

Fig. 6 Cut-away schematic view of a CMOS inverter. (From Maly, [11]. Reprinted with permission, Addison-Wesley Publishing Co.)

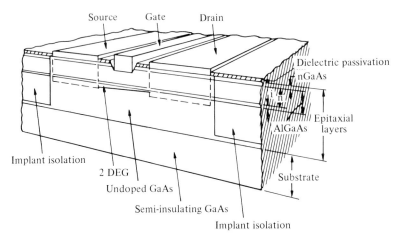

Fig. 7 Perspective view of a MODFET structure similar to those used in prototype integrated circuits. (From Drummond et al., [12], © 1986 IEEE.)

reader to relevant terms, concepts, models, and analytical procedures. Chapter 1 is primarily devoted to the junction field effect transistor (J-FET), although the discussion and analysis is equally applicable to the closely related MESFET (metal-semiconductor field-effect transistor). Since most readers will already be familiar with *pn* junction operation, the J-FET provides a conceptual bridge between the *pn* junction devices and the "pure" field-effect devices considered later in the volume. The J-FET discussion, moreover, serves as a first run through on terminology, analytical procedures, and the like, thereby making the subsequent discussion easier to understand. It should be noted, however, that the latter chapters are self-contained, with no specific references to the J-FET/MESFET presentation. Chapters 2 and 3 provide a base-line coverage of MOS structural and device fundamentals. The two-terminal MOS-capacitor or MOS-C is the structural heart of all MOS devices. In Chapter 2 we define the ideal MOS-C structure, present a qualitative and quantitative description of the structure in the static state, and examine the capacitance-voltage characteristics derived from the device. The operational description and analysis of the basic transistor configuration, the long-channel enhancement-mode MOSFET, are presented in Chapter 3. Finally, Chapters 4 and 5 contain MOS-related practical information. The nature and effect of deviations from the ideal are described in Chapter 4, while Chapter 5 provides an introductory discussion of small-dimension effects in MOSFETs. A brief survey of modified and developmental FET structures is also included in Chapter 5.

After completing the present volume, the reader interested in learning about CCDs and the MOS memory cell mentioned earlier in the General Introduction, or just learning more about MOS devices in general, is referred to Volume VII in the Modular Series, *Advanced MOS Devices,* by D. K. Schroder. Volume V in the series, *Introduction to Microelectronic Fabrication,* by R. C. Jaeger, should be consulted for MOS fabrication details. Lastly, a greatly expanded and highly detailed review of MOS device history can be found in C. T. Sah's article [8].

REFERENCES

[1] W. Shockley, "A Unipolar Field-Effect Transistor," *Proc. IRE,* **40**, 1365 (November 1952).

[2] G. C. Dacey and I. M. Ross, "Unipolar Field-Effect Transistor," *Proc. IRE,* **41**, 970 (August 1953).

[3] G. C. Dacey and I. M. Ross, "The Field-Effect Transistor," *Bell System Tech. J.,* **34**, 1149 (November 1955).

[4] D. Kahng and M. M. Atalla, "Silicon-Silicon Dioxide Field Induced Surface Devices," presented at the IRE-AIEE Solid-State Device Research Conference, Carnegie Institute of Technology, Pittsburgh, Pa., 1960.

[5] J. S. MacDougall, "Applications of the Silicon Planar II. MOSFET," *Application Bulletin,* Fairchild Semiconductor, November, 1964.

[6] R. H. Dennard, "Field-Effect Transistor Memory," U.S. Patent 3 387 286, application filed July 14, 1967, granted June 4, 1968.

[7] P. Wyns and R. L. Anderson, "Low-Temperature Operation of Silicon Dynamic Random-Access Memories," *IEEE Trans. on Electron Devices,* **36**, 1423 (August 1989).

[8] C. T. Sah, "Evolution of the MOS Transistor — From Conception to VLSI," *Proc. IEEE,* **76**, 1280 (October 1988).

[9] M. F. Tompsett, G. F. Amelio, and G. E. Smith, "Charge Coupled 8-Bit Shift Register," *Appl. Phys. Lett.,* **17**, 111 (1970).

[10] F. M. Wanlass and C. T. Sah, "Nanowatt Logic Using Field-effect Metal-Oxide-Semiconductor Triodes," in *Technical Digest of IEEE 1963 Int. Solid-State Circuit Conf.,* pp. 32–33, February 20, 1963. F. M. Wanlass, "Low Standy-By Power Complementary Field-Effect Circuitry," U.S. Patent 3 356 858 filed June 18, 1963, issued December 5, 1967.

[11] W. Maly, *Atlas of IC Technologies: An Introduction to VLSI Processes,* The Benjamin/ Cummings Publishing Co., Inc., Menlo Park, Calif., © 1987; p. 191.

[12] T. J. Drummond, W. T. Masselink, and H. Morkoç, "Modulation-Doped GaAs/(Al, Ga)As Heterojunction Field-Effect Transistors: MODFETs," *Proc. IEEE,* **74**, 773 (June 1986).

1 / J-FET and MESFET

1.1 INTRODUCTION

As mentioned in the General Introduction, the junction field effect transistor (J-FET) was first proposed and analyzed by W. Shockley in 1952. A cross section of the basic device structure is shown in Fig. 1.1. In the J-FET the application of a gate voltage varies the *pn*-junction depletion widths and the associated electric field in the direction normal to the semiconductor surface. Changes in the depletion widths, in turn, modulate the conductance between the ohmic source and drain contacts.

The J-FET was initially named the unipolar transistor to distinguish it from the bipolar junction transistor and to emphasize that only one type of carrier was involved in the operation of the new device. Specifically, for the structure pictured in Fig. 1.1, normal operation of the transistor can be described totally in terms of the electrons flowing in the *n*-region from the source to the drain. The source (S) terminal gets its name from the fact that the carriers contributing to the current flow move from the external circuit into the semiconductor at this electrode. The carriers leave the semiconductor, or are "drained" from the semiconductor, at the drain (D) electrode. The gate is so named because of its control or gating action. The modern version of the J-FET

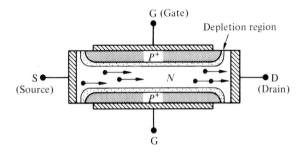

Fig. 1.1 Cross section of the basic J-FET structure.

shown in Fig. 1.2, although somewhat different in physical appearance, is functionally equivalent to the original Shockley structure.

Rectifying metal-semiconductor (or Schottky barrier) diodes and *pn* junction diodes are closely related in many respects (see Modular Series Volume II, Chapter 7). Notably, a depletion region, which can be modulated by the applied voltage, also forms beneath a rectifying metal-semiconductor contact. It logically follows, as proposed by Mead in 1966, that field-effect transistors can be built with rectifying metal-semiconductor gates. Idealized and modern versions of the metal-semiconductor field-effect transistor (MESFET) are pictured in Fig. 1.3(a) and 1.3(b), respectively. As might be inferred from Fig. 1.3(b), MESFETs are almost exclusively encountered in GaAs-based circuits. The use of MESFETs in GaAs integrated circuits is due to the lack of a metal-insulator-GaAs technology and the immaturity of the promising modulation-doped (MODFET) technology. In the discussion and analysis to follow, we make sole reference to the J-FET. It should be understood, however, that the arguments and results likewise apply with minor modifications to large-dimension MESFETs.

1.2 QUALITATIVE THEORY OF OPERATION

To establish the basic principles of J-FET operation, we will assume standard biasing conditions and treat the symmetrical, somewhat idealized, Shockley structure of Fig. 1.1. Given an *n*-type region between the source and drain, standard operational

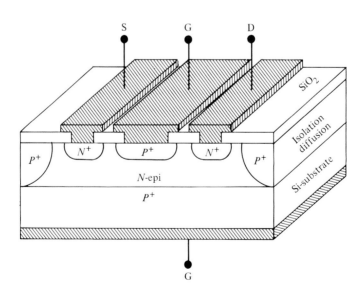

Fig. 1.2 Perspective view of a modern epilayer J-FET.

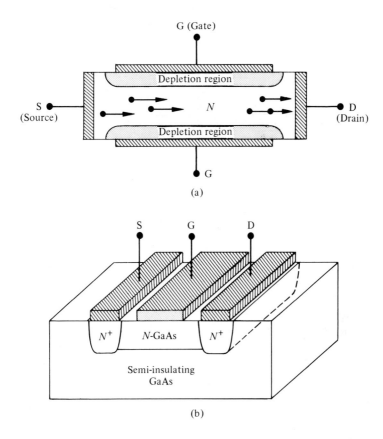

Fig. 1.3 The MESFET. (a) Idealized structure and (b) simplified perspective view of a modern GaAs MESFET.

conditions prevail in the J-FET when the top and bottom gates are tied together, $V_G \leq 0$, and $V_D \geq 0$, as illustrated in Fig. 1.4. Note that with $V_G \leq 0$, the *pn* junctions are always zero or reverse biased. Also, $V_D \geq 0$ ensures that the electrons in the *n*-region move from the source to the drain (in agreement with the naming of the S and D terminals). Our approach here will be to systematically change the terminal voltages and examine what is happening inside the device.

First suppose that the gate terminal is grounded, $V_G = 0$, and the drain voltage is increased in small steps starting from $V_D = 0$. At $V_D = 0$ (remember V_G is also zero) the device is in thermal equilibrium and about all one sees inside the structure are small depletion regions about the top and bottom p^+-n junctions [see Fig. 1.5(a)]. The depletion regions extend, of course, primarily into the lightly doped, central *n*-region of the device. Stepping V_D to small positive voltages yields the situation pictured in Fig. 1.5(b). A current, I_D, begins to flow into the drain and through the nondepleted

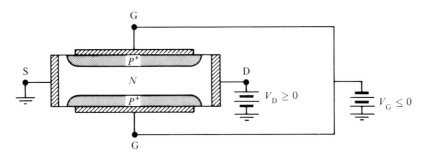

Fig. 1.4 Specification of the device structure and biasing conditions assumed in the qualitative analysis.

n-region sandwiched between the two p^+-n junctions. The nondepleted, current-carrying region, we might note, is referred to as the *channel*. For small V_D, the channel looks and acts like a simple resistor, and the resulting variation of I_D with V_D is linear [see Fig. 1.6(a)].

When V_D is increased above a few tenths of a volt, the device typically enters a new phase of operation. To gain insight into the revised situation, refer to Fig. 1.5(c), where an arbitrarily chosen potential of 5 V is assumed to exist at the drain terminal. Since the source is grounded, it naturally follows that somewhere in the channel the potential takes on the values of 1, 2, 3, and 4 volts, with the potential increasing as one progresses from the source to the drain. The p^+ sides of the p^+-n junctions, however, are being held at zero bias. Consequently, the bias applied to the drain leads indirectly to a reverse biasing of the gate junctions and an increase in the junction depletion widths. Moreover, the top and bottom depletion regions progressively widen in going down the channel from the source to the drain [see Fig. 1.5(d)]. Still thinking of the channel region (the nondepleted n-region) as a resistor, but no longer a simple resistor, one would expect the loss of conductive volume to increase the source-to-drain resistance and reduce the ΔI_D resulting from a given change in drain voltage. This is precisely the situation pictured in Fig. 1.6(b). The slope of the I_D–V_D characteristic decreases at larger drain biases because of the channel-narrowing effect.

Continuing to increase the drain voltage obviously causes the channel to narrow more and more, especially near the drain, until eventually the top and bottom depletion regions touch in the near vicinity of the drain, as pictured in Fig. 1.5(e). The complete depletion of the channel, touching of the top and bottom depletion regions, is an important special condition and is referred to as "*pinch-off*." When the channel pinches off inside the device, the slope of the I_D–V_D characteristic becomes approximately zero [see Fig. 1.6(c)], and the drain bias at the pinch-off point is given the special designation V_{Dsat}. For drain biases in excess of V_{Dsat}, the I_D–V_D characteristic saturates, that is, remains approximately constant at the I_{Dsat} value.

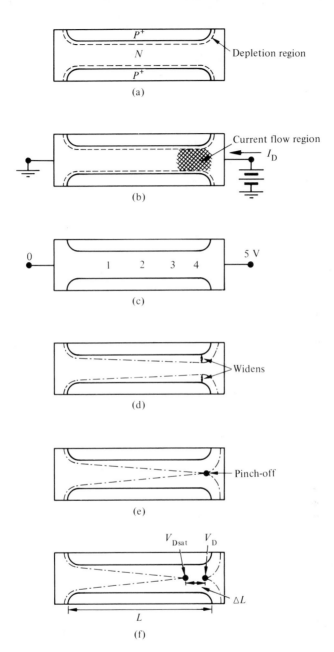

Fig. 1.5 Visualization of various phases of $V_G = 0$ J-FET operation. (a) Equilibrium ($V_D = 0$, $V_G = 0$); (b) small V_D biasing; (c) voltage drop down the channel for an arbitrarily assumed $V_D = 5$V; (d) channel narrowing under moderate V_D biasing; (e) pinch-off; (f) postpinch-off ($V_D > V_{Dsat}$).

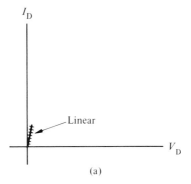

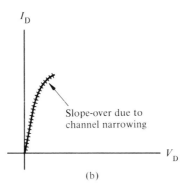

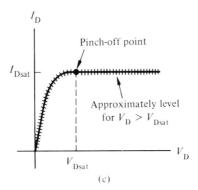

Fig. 1.6   General form of the I_D–V_D characteristics. (a) Linear, simple resistor, variation for very small drain voltages. (b) Slope-over at moderate drain biases due to channel narrowing. (c) Pinch-off and saturation for drain voltages in excess of V_{Dsat}.

The statements presented without explanation in the preceding paragraph are totally factual. The I_D–V_D characteristic does level off or saturate when the channel pinches off. At first glance, however, the facts appear to run contrary to physical intuition. Should not pinch-off totally eliminate any current flow in the channel? How can one account for the fact that V_D voltages in excess of V_{Dsat} have essentially no effect on the drain current?

In answer to the first question, let us suppose I_D at pinch-off was identically zero. If I_D were zero, there would be no current in the channel at any point and the voltage down the channel would be the same as at $V_D = 0$, namely, zero everywhere. If the channel potential is zero everywhere, the *pn* junctions would be zero biased and the channel in turn would be completely open from the source to the drain, clearly contradicting the initial assumption of a pinched-off channel. In other words, a current must flow in the J-FET to induce and maintain the pinched-off condition. Perhaps the conceptual difficulty often encountered with pinch-off arises from the need for a large current to flow through a depletion region. Remember, depletion regions are not totally devoid of carriers. Rather, the carrier numbers are just small compared to the background doping concentration (N_D or N_A) and may still approach densities $\sim 10^{12}/cm^3$ or greater. Moreover, the passage of large currents through a depletion region is not unusual in solid-state devices. For example, a large current flows through the depletion region in a forward biased diode and through both depletion regions in a bipolar junction transistor.

With regard to the saturation of I_D for drain biases in excess of V_{Dsat}, there is a very simple physical explanation. When the drain bias is increased above V_{Dsat} the pinched-off portion of the channel widens from just a point into a depleted channel section ΔL in extent. As shown in Fig. 1.5(f), the voltage on the drain side of the ΔL section is V_D, while the voltage on the source side of the section is V_{Dsat}. In other words, the applied drain voltage in excess of V_{Dsat}, $V_D - V_{Dsat}$, is dropped across the depleted section of the channel. Now, assuming $\Delta L \ll L$, the usual case, the source-to-pinch-off region of the device will be essentially identical in shape and will have the same endpoint voltages (zero and V_{Dsat}) as were present at the start of saturation. If the shape of a conducting region and the potential applied across the region do not change, then the current through the region must also remain invariant. This explains the approximate constancy of the drain current for postpinch-off biasing. [Naturally, if ΔL is comparable to L, then the same voltage drop (V_{Dsat}) will appear across a shorter channel section ($L - \Delta L$) and the postpinch-off I_D will increase perceptibly with increasing $V_D > V_{Dsat}$. This effect is especially noticeable in short channel (small L) devices.]

Another approach to explaining the saturation of the I_D–V_D characteristics makes use of an analogous situation in everyday life; namely, a waterfall. As everyone knows, the water-flow rate over a falls is controlled not by the height of the falls, but by the flow rate down the rapids leading to the falls. Thus, assuming an identical rapids region, the water-flow rate at the bottom of the two falls pictured in Fig. 1.7 is precisely the same, even though the heights of the falls are different. The rapids region is of course analogous to the source side of the channel in the J-FET, the falls proper corresponds to the pinched-off ΔL section at the drain end of the channel, and the height of the falls corresponds to the $V_D - V_{Dsat}$ potential drop across the ΔL section.

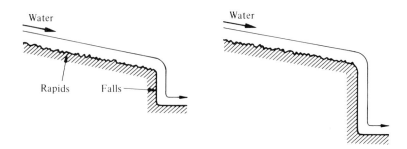

Fig. 1.7   The waterfalls analogy.

Thus far we have established the expected variation of I_D with V_D when $V_G = 0$. To complete the discussion we need to investigate the operation of the J-FET when $V_G < 0$. It turns out that $V_G < 0$ operation is very similar to $V_G = 0$ operation with three minor modifications. First, if $V_G < 0$ the top and bottom p^+-n junctions are reverse biased even when $V_D = 0$. A reverse bias on the junctions increases the width of the depletion regions and shrinks the $V_D = 0$ lateral width of the channel. Consequently, the resistance of the channel increases at a given V_D value and the linear portion of the I_D-V_D characteristic exhibits a smaller slope when $V_G < 0$ [see Fig. 1.8(a)]. Second, because the channel is narrower at $V_D = 0$, the channel also becomes pinched-off at a smaller drain bias. Therefore, as pictured in Fig. 1.8(b), V_{Dsat} and I_{Dsat} when $V_G < 0$ are smaller than V_{Dsat} and I_{Dsat} when $V_G = 0$. Finally, note that for sufficiently negative V_G biases it is possible to deplete the entire channel even with $V_D = 0$ [see Fig. 1.8(c)]. The gate bias, $V_G = V_P$, where the gate voltage first totally depletes the entire channel with V_D set equal to zero, is referred to as the pinch-off gate voltage. For $V_G \leq V_P$ the drain current is identically zero for all drain biases.*

1.3   QUANTITATIVE I_D-V_D RELATIONSHIPS

Wanted: a quantitative expression for the drain current as a function of the terminal voltages; that is, $I_D = I_D(V_D, V_G)$.

Device Specification. The precise device structure, dimensions, and assumed coordinate orientations are as specified in Fig. 1.9. The y-axis is directed down the channel from the source to the drain while the x-coordinate is oriented normal to the p^+-n metallurgical junctions, L is the channel length, Z is the p^+-n junction width, and $2a$ is the distance between the top and bottom metallurgical junctions. Note that $y = 0$ and $y = L$ are slightly removed from the source and drain contacts, respectively. $V(y)$ is the po-

---

*If the drain bias is made very large, the p^+-n junctions in the vicinity of the drain will eventually break down, leading to a very rapid increase in I_D with V_D for any gate bias. This breakdown has been omitted from all of the theoretical characteristics sketched herein.

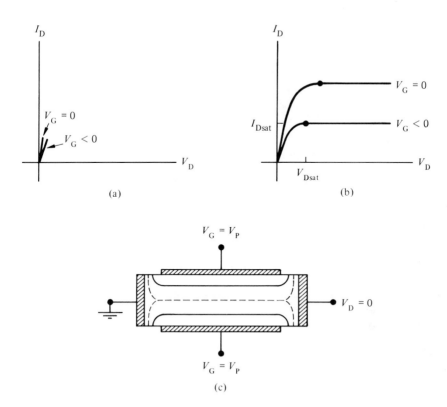

Fig. 1.8 Modification of the I_D–V_D characteristics when $V_G < 0$. (a) Decrease in the linear slope of the characteristics for small drain voltages. (b) Decrease in the saturation current and saturation drain voltage. (c) Gate pinch-off.

tential and $W(y) = W_{top}(y) = W_{bottom}(y)$ is the junction depletion width at an aribtrary point y in the channel. $W(y)$ lies, of course, almost totally in the n-region because of the p^+-n nature of the junctions.

Basic Assumptions. (1) The junctions are p^+-n *step* junctions and the n-region is uniformly doped with a donor concentration equal to N_D. (2) The device is structurally symmetrical about the $x = a$ plane as shown in Fig. 1.9, and the symmetry is maintained by operating the device with the same V_G bias applied to the top and bottom gates. (3) Current flow is confined to the nondepleted portions of the n-region and is directed exclusively in the y-direction. (4) $W(y)$ can be increased to at least a width a without inducing breakdown in the p^+-n junctions. (We implicitly assumed this to be the case in the qualitative discussion.) (5) Voltage drops from the source to $y = 0$ and from $y = L$ to the drain are negligible. (6) $L \gg a$.

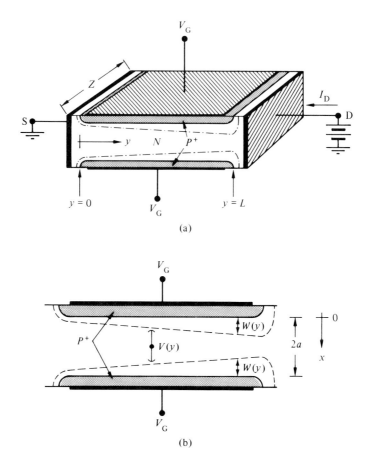

Fig. 1.9 Device structure, dimensions, and coordinate orientations assumed in the quantitative analysis. (a) Overall diagram. (b) Expanded view of the channel region.

For drain and gate voltages below pinch-off, $0 \leq V_D \leq V_{Dsat}$ and $0 \geq V_G \geq V_P$, the derivation of the desired I_D–V_D relationship proceeds as follows: In general one can write

$$\mathbf{J_N} = q\mu_n n \boldsymbol{\mathscr{E}} + qD_N \nabla n \qquad (1.1)$$

Within the conducting channel $n \simeq N_D$ and the current is flowing almost exclusively in the y-direction. Moreover, with $n \simeq N_D$, the diffusion component of the current ($qD_N \nabla n$) should be relatively small. Under the cited conditions Eq. (1.1) reduces to

$$J_N = J_{Ny} = q\mu_n N_D \mathscr{E}_y = -q\mu_n N_D \frac{dV}{dy} \qquad \text{(in the conducting channel)} \qquad (1.2)$$

Since there are no carrier sinks or sources in the device, the current flowing through any cross-sectional plane within the channel must be equal to I_D. Thus, integrating the current density over the cross-sectional area of the conducting channel at an arbitrary point y yields

$$I_D = -\int\int J_{Ny}\,dx\,dz = -Z\int_{W(y)}^{2a-W(y)} J_{Ny}\,dx = 2Z\int_{W(y)}^{a} q\mu_n N_D \frac{dV}{dy}\,dx \qquad (1.3a)$$

$$= 2qZ\mu_n N_D a\frac{dV}{dy}\left(1 - \frac{W}{a}\right) \qquad (1.3b)$$

The minus sign appears in the general formula for I_D because I_D is defined to be positive in the $-y$ direction. Use was also made of the fact that the structure is symmetrical about the $x = a$ plane.

Remembering that I_D is independent of y, one can recast Eq. (1.3b) into a more useful form by integrating I_D over the length of the channel. Specifically,

$$\int_0^L I_D\,dy = I_D L = 2qZ\mu_n N_D a\int_{V(0)=0}^{V(L)=V_D}\left[1 - \frac{W(V)}{a}\right]dV \qquad (1.4)$$

or

$$I_D = \frac{2qZ\mu_n N_D a}{L}\int_0^{V_D}\left[1 - \frac{W(V)}{a}\right]dV \qquad (1.5)$$

To proceed any further we need an analytical expression for W as a function of V. It should be recognized that the electrostatic problem inside the J-FET is really two-dimensional in nature. To obtain an exact expression for W as a function of V would necessitate the solution of Poisson's equation taking into account both the x and y variation of the electrostatic variables. Fortunately, with $L \gg a$ (basic assumption 6), the y-direction dependence of the electrostatic variables (potential, electric field, etc.) in the J-FET is expected to be a much more gradual function of position than the x-direction dependence of the same variables. This allows us to invoke the *gradual channel approximation,* an approximation that is encountered frequently in the analysis of field-effect devices. Specifically, the "down-the-channel" or y-dependence is assumed to have little effect on the x-direction computation of the electrostatic variables, thereby permitting a pseudo-one-dimensional analysis in the x-direction.

For the problem under consideration, invoking the gradual channel approximation simply means that we can approximate W at every point y using the one-dimensional expression established in Modular Series Volume II (see Volume II, Section 2.4).

Thus, for the assumed p^+-n step junctions,

$$W(V) \cong \left[\frac{2K_S \varepsilon_0}{qN_D}(V_{bi} - V_A)\right]^{1/2} = \left[\frac{2K_S \varepsilon_0}{qN_D}(V_{bi} + V - V_G)\right]^{1/2} \qquad (1.6)$$

where, as is evident from Fig. 1.9(b), $V_A = V_G - V(y)$ is the applied potential drop across the junction at a given point y. It is next convenient to note that $W \to a$ when $V_D = 0$ ($V = 0$) and $V_G = V_P$. Thus substituting into Eq. (1.6) yields

$$a = \left[\frac{2K_S \varepsilon_0}{qN_D}(V_{bi} - V_P)\right]^{1/2} \qquad (1.7)$$

and

$$\frac{W(V)}{a} = \left(\frac{V_{bi} + V - V_G}{V_{bi} - V_P}\right)^{1/2} \qquad (1.8)$$

Finally, substituting $W(V)/a$ from Eq. (1.8) into Eq. (1.5) and performing the indicated integration, one obtains

$$I_D = \frac{2qZ\mu_n N_D a}{L}\left\{V_D - \frac{2}{3}(V_{bi} - V_P)\left[\left(\frac{V_D + V_{bi} - V_G}{V_{bi} - V_P}\right)^{3/2} - \left(\frac{V_{bi} - V_G}{V_{bi} - V_P}\right)^{3/2}\right]\right\} \qquad (1.9)$$
$$\text{for } 0 \le V_D \le V_{Dsat}; \qquad V_P \le V_G \le 0$$

Equation (1.9) could be simplified by introducing $G_0 \equiv 2qZ\mu_n N_D a/L$; physically, G_0 is the channel conductance one would observe if there were no depletion regions. We have retained $2qZ\mu_n N_D a/L$ in the text expressions so that the major parametic dependencies are immediately obvious.

It should be reemphasized that the foregoing development and Eq. (1.9), in particular, apply only below pinch-off. In fact, the computed I_D versus V_D for a given V_G actually begins to decrease if V_D values in excess of V_{Dsat} are inadvertently substituted into Eq. (1.9). As pointed out in the qualitative discussion, however, I_D is approximately constant if V_D exceeds V_{Dsat}. To first order, then, the postpinch-off portion of the characteristics can be modeled by simply setting

$$I_{D|V_D>V_{Dsat}} = I_{D|V_D=V_{Dsat}} \equiv I_{Dsat} \qquad (1.10a)$$

or

$$I_{Dsat} = \frac{2qZ\mu_n N_D a}{L}\left\{V_{Dsat} - \frac{2}{3}(V_{bi} - V_P)\left[\left(\frac{V_{Dsat} + V_{bi} - V_G}{V_{bi} - V_P}\right)^{3/2} - \left(\frac{V_{bi} - V_G}{V_{bi} - V_P}\right)^{3/2}\right]\right\}$$

(1.10b)

The I_{Dsat} relationship can be simplified somewhat by noting that pinch-off at the drain end of the channel implies $W \to a$ when $V(L) = V_{Dsat}$. Therefore, from Eq. (1.6),

$$a = \left[\frac{2K_S\varepsilon_0}{qN_D}(V_{bi} + V_{Dsat} - V_G)\right]^{1/2}$$

(1.11)

Comparing Eqs. (1.7) and (1.11) one concludes

$$\boxed{V_{Dsat} = V_G - V_P}$$

(1.12)

and

$$\boxed{I_{Dsat} = \frac{2qZ\mu_n N_D a}{L}\left\{V_G - V_P - \frac{2}{3}(V_{bi} - V_P)\left[1 - \left(\frac{V_{bi} - V_G}{V_{bi} - V_P}\right)^{3/2}\right]\right\}}$$

(1.13)

Theoretical I_D–V_D characteristics computed using Eqs. (1.9) and (1.13) are presented in Fig. 1.10. For comparison purposes a sample set of experimental characteristics are displayed in Fig. 1.11. Generally speaking, the theory does a reasonably adequate job of modeling the experimental observations. A somewhat improved agreement between experiment and theory can be achieved by lifting the assumption of negligible voltage drops in the regions of the device between the active channel and the source/drain contacts (see Fig. 1.12). We leave it to the reader as an exercise to establish the resulting theoretical modifications. Finally, it should be pointed out that in saturation most J-FET characteristics can be closely modeled by the simple relationship

$$\boxed{I_{Dsat} = I_{D0}(1 - V_G/V_P)^2}\qquad \text{where } I_{D0} = I_{Dsat|V_G=0}$$

(1.14)

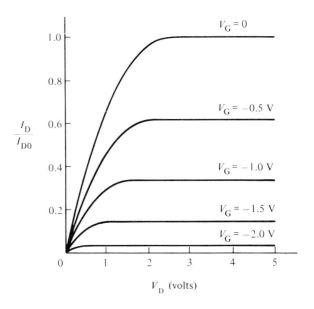

Fig. 1.10   Normalized theoretical I_D–V_D characteristics assuming $V_{bi} = 1$ V and $V_P = -2.5$ V. $I_{D0} = I_{Dsat|V_G=0}$.

Although appearing totally different than Eq. (1.13), the semiempirical "square-law" relationship of Eq. (1.14) yields similar numerical results and is much easier to use in performing first order circuit calculations where the J-FET is viewed as a "black-box." Equation (1.13), on the other hand, is indispensable if one wishes to investigate the dependence of the J-FET characteristics on temperature, channel doping, or some other fundamental device parameter.

SEE EXERCISE 1.1 — APPENDIX A

1.4 ac RESPONSE

The ac response of the J-FET, routinely expressed in terms of the J-FET's small signal equivalent circuit, is most conveniently established by considering the two-port network shown in Fig. 1.13. Herein we will restrict our considerations to low operational frequencies where capacitive effects may be neglected.

Let us begin by examining the device input. Under standard dc biasing conditions, the input port between the gate and source is connected across a reverse biased diode on the inside of the structure. A reverse biased diode, however, behaves (to first order) like an open circuit at low frequencies. It is standard practice, therefore, to model the input to the J-FET by an open circuit.

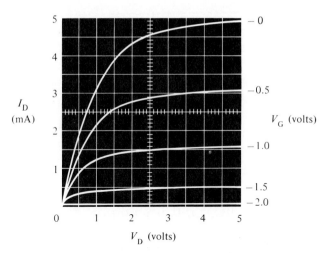

Fig. 1.11 Sample experimental I_D–V_D characteristics. (Characteristics were derived from a TI 2N3823 n-channel J-FET.)

At the output port, the dc drain current has already been established to be a function of V_D and V_G; that is, $I_D = I_D(V_D, V_G)$. When the ac drain and gate potentials, v_d and v_g, are respectively added to the dc drain and gate terminal voltages, V_D and V_G, the drain current through the structure is of course modified to $I_D(V_D, V_G) + i_d$, where i_d is the ac component of the drain current. Provided the device can follow the ac changes in potential quasi-statically,* which is assumed to be the case at low operational frequencies, one can state

$$i_d + I_D(V_D, V_G) = I_D(V_D + v_d, V_G + v_g) \tag{1.15a}$$

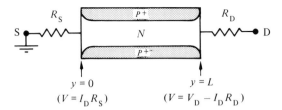

Fig. 1.12 Modified model for the J-FET including the resistance of the semiconductor regions between the ends of the active channel and the source/drain terminals.

---

*The term "quasi-static" is used to describe situations where the time varying state of a system at any given instant is essentially indistinguishable from the dc state that would be achieved under equivalent biasing conditions.

Fig. 1.13   The J-FET viewed as a two-port network.

and

$$i_d = I_D(V_D + v_d, V_G + v_g) - I_D(V_D, V_G) \tag{1.15b}$$

Expanding the first term on the right-hand side of Eq. (1.15b) in a Taylor series and keeping only first-order terms in the expansion (higher-order terms are negligible), one obtains

$$i_d = \left.\frac{\partial I_D}{\partial V_D}\right|_{V_G} v_d + \left.\frac{\partial I_D}{\partial V_G}\right|_{V_D} v_g \tag{1.16a}$$

or

$$i_d = g_d v_d + g_m v_g \tag{1.16b}$$

where

$g_d \equiv \left.\dfrac{\partial I_D}{\partial V_D}\right|_{V_G=\text{constant}}$	the drain or channel conductance	(1.17a)
$g_m \equiv \left.\dfrac{\partial I_D}{\partial V_G}\right|_{V_D=\text{constant}}$	transconductance or mutual conductance	(1.17b)

Equation (1.16b) may be viewed as the ac-current node equation for the drain terminal and, by inspection, leads to the output portion of the circuit displayed in Fig. 1.14. Since, as concluded earlier, the gate-to-source or input portion of the device is simply an open circuit, Fig. 1.14 is the desired small signal equivalent circuit characterizing the low-frequency ac response of the J-FET. Note that for field-effect transistors the g_m parameter plays a role analogous to the α's and β's in the modeling of bipolar junction transistors. As its name indicates, g_d may be viewed as either the

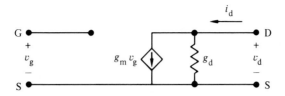

Fig. 1.14 Small signal equivalent circuit characterizing the low-frequency ac response of the J-FET.

device output admittance or the ac conductance of the channel between the source and drain. Explicit g_d and g_m relationships obtained by direct differentiation of Eqs. (1.9) and (1.13), using the Eq. (1.17) definitions, are catalogued in Table 1.1. The $g_d = 0$ result in Table 1.1 for operation of the device under saturation conditions is of course consistent with the theoretically zero slope of the I_D-V_D characteristics when $V_D \geq V_{Dsat}$.

SEE EXERCISE 1.2 — APPENDIX A

PROBLEMS

1.1 Answer the following questions as concisely as possible.

(a) Define "field-effect."

(b) Precisely what is the "channel" in J-FET terminology?

(c) What is meant by the term "pinch-off"?

(d) For a p-channel device (a device with n^+-p gating junctions and a p-region between the source and drain), does the drain current flow into or out of the drain contact under normal operational conditions? Explain.

(e) What is the "gradual channel approximation"?

Table 1.1 J-FET Small Signal Parameters. Entries in the table were obtained by direct differentiation of Eqs. (1.9) and (1.13). $G_0 \equiv 2qZ\mu_n N_D a/L$.

Below pinch-off ($V_D \leq V_{Dsat}$)	Post-pinch-off ($V_D \geq V_{Dsat}$)
$g_d = G_0 \left[1 - \left(\dfrac{V_D + V_{bi} - V_G}{V_{bi} - V_P} \right)^{1/2} \right]$	$g_d = 0$
$g_m = G_0 \left[\left(\dfrac{V_D + V_{bi} - V_G}{V_{bi} - V_P} \right)^{1/2} - \left(\dfrac{V_{bi} - V_G}{V_{bi} - V_P} \right)^{1/2} \right]$	$g_m = G_0 \left[1 - \left(\dfrac{V_{bi} - V_G}{V_{bi} - V_P} \right)^{1/2} \right]$

(f) What is the mathematical definition of the drain conductance? of the transconductance?

(g) Draw the small signal equivalent circuit characterizing the low-frequency ac response of a *long*-channel J-FET under *saturation* conditions. (Assume the dc characteristics of the device are similar to those shown in Fig. 1.10.)

1.2 Consider the MESFET structure shown in Fig. 1.3(b). Although a depletion region exists adjacent to the semi-insulating substrate, the depletion width is small and only weakly modulated by the applied drain and gate voltages. In effect, the MESFET is a one-sided structure. If the width of the *n*-GaAs is taken to be a, how must the development in Section 1.3 be modified to account for the one-sided nature of the MESFET structure?

1.3 If Eq. (1.9) is used to compute I_D as a function of V_D for a given V_G, and if V_D is allowed to increase above V_{Dsat}, one finds I_D to be a peaked function of V_D maximizing at V_{Dsat}. The foregoing suggests a second way to establish the Eq. (1.12) relationship for V_{Dsat}. Specifically, show that the standard mathematical procedure for determining extrema points of a function can be used to derive Eq. (1.12) directly from Eq. (1.9).

1.4 Compute I_{Dsat}/I_{D0} from the square-law relationship [Eq. (1.14)] assuming $V_P = -2.5$ V and $V_G = 0$ V, -0.5 V, -1.0 V, -1.5 V, -2.0 V. Compare your results with the Eq. (1.13) based I_{Dsat}/I_{D0} values shown in Fig. 1.10. Comment on the comparison.

1.5 The maximum frequency of operation or cutoff frequency of the J-FET is given by

$$f_{max} = \frac{g_m}{2\pi C_G}$$

where C_G is the capacitance of the *pn*-junction gates.

(a) Derive the above expression for f_{max} by presenting an argument analogous to that found in Subsection 3.3.2.

(b) Show that for a J-FET one can write

$$f_{max} = \frac{g_m}{2\pi C_G} \leq \frac{q\mu_n N_D a^2}{2\pi K_s \varepsilon_o L^2}$$

(c) Given a silicon J-FET with $N_D = 10^{16}/\text{cm}^3$, $a = 0.5$ μm and $L = 5$ μm, compute the limiting value of the cutoff frequency.

1.6 In this problem we wish to explore the temperature dependence of the J-FET transconductance.

(a) Making free use of plots in Volume I of the Modular Series, compute $g_m(T)$ normalized to $G_0(300$ K) at 50° C intervals between $-50°$ C and 200° C. Set $V_G = 0$ and take the device to be saturation biased ($V_D \geq V_{Dsat}$). Assume an *n*-channel silicon device with $N_D = 10^{16}/\text{cm}^3$, N_A(of the P^+ regions) $= 5 \times 10^{17}/\text{cm}^3$, and $a = 0.6$ μm. [*Note*: Neglecting the very small change in device dimensions with temperature, we conclude from Eq. (1.7) that $V_{bi} - V_P$ must be temperature independent. Separately, however, V_{bi} and V_P *are* temperature dependent.]

(b) Plot the computed $g_m(T)/G_0(300\ \text{K})$ values versus T(in kelvin) on log-log graph paper. Also plot $\mu_n(T)/\mu_n(300\ \text{K})$ on the same set of coordinates. Assuming $g_m/G_0 \propto T(\text{K})^{-n}$, determine **n**. Briefly discuss your results.

1.7 Bipolar integrated circuits occasionally make use of *pinch resistors*. The two-terminal pinch resistor is essentially a J-FET with the *gate(s) internally shorted to the source*. Exhibiting a voltage-dependent nonlinear resistance, the device finds use in applications that require large values of resistance, but where the precise values are not critical.

(a) Assuming the text development can be used without modification, establish general expressions for the dc conductance ($G = 1/\text{Resistance} = I/V$) and the ac conductance ($g = dI/dV$) of the pinch resistor.

(b) Compute the value of the dc and ac resistances ($R = 1/G$ and $r = 1/g$) at an applied voltage of $V_{\text{Dsat}}/2$. Employ $Z/L = 1$, $a = 0.5\ \mu\text{m}$, $N_D = 10^{16}/\text{cm}^3$, $V_{\text{bi}} = 1\ \text{V}$, and $V_P = -2\ \text{V}$.

1.8 Suppose, as shown in Fig. P1.8, the bottom gate lead of a J-FET is tied to the source and grounded.

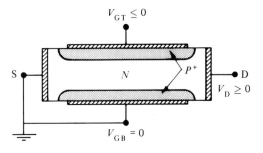

Fig. P1.8

(a) Sketch an outline of the depletion regions inside the $V_{\text{GB}} = 0$ device when V_{GT} is made sufficiently large to pinch-off the channel with $V_D = 0$.

(b) If V_P (the pinch-off gate voltage) $= -8\ \text{V}$ when the two gates are tied together, $V_{\text{bi}} = 1\ \text{V}$, and assuming p^+-n step junctions, determine V_{PT} (the top gate pinch-off voltage) when $V_{\text{GB}} = 0$ and $V_D = 0$. Is your answer here consistent with the sketch in part (a)? Explain.

(c) Assuming $V_{\text{PT}} < V_{\text{GT}} < 0$, sketch an outline of the depletion regions inside the device when the *drain* voltage is increased to the pinch-off point.

(d) Derive an expression which specifies V_{Dsat} in terms of V_{PT}, V_{bi}, and V_{GT} for $V_{\text{GB}} = 0$ operation. (Your answer should contain only voltages. Make no attempt to actually solve for V_{Dsat}.)

(e) In light of your answers to parts (c) and (d), will V_{Dsat} for $V_{\text{GB}} = 0$ operation be greater than or less than the V_{Dsat} for $V_{\text{GB}} = V_{\text{GT}}$ operation? Explain.

(f) Derive an expression for I_D as a function of V_D and V_{GT} analogous to Eq. (1.9).

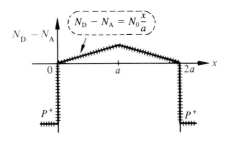

Fig. P1.9

1.9 A J-FET is constructed with the *gate-to-gate* doping profile shown in Fig. P1.9. Specifically assume the p^+-region doping is much greater than the maximum n-region doping and make other obvious assumptions as required.

(a) Starting with Poisson's equation, derive an expression for the *left-hand* junction depletion width (W). Let V_A be the applied voltage drop across the junction. (If necessary, refer to Section 2.5 in Volume II of the Modular Series.)

(b) Neglecting the μ_n doping dependence and assuming the left-hand and right-hand gates are tied together, appropriately modify the text J-FET analysis to obtain the below-pinch-off I_D–V_D relationship for this linearly graded junction. (*Caution:* more than the $W(V)/a$ expression must be modified.)

1.10 Referring to Exercise 1.1 in Appendix A, and permitting arbitrary values of R_S and R_D, write a microcomputer program (possibly using Student MathCAD available from Addison-Wesley Publishing Co.) that can be used to compute and plot J-FET I_D–V_D characteristics. Normalize all I_D values to $I_{D0} \equiv I_{Dsat|V_G=0}$ *as computed from Eq. (1.13)*. (The normalized current relationships should involve only I_D/I_{D0} or I_{Dsat}/I_{D0}, V_D, V_G, and the parameters V_{bi}, V_P, $G_0 R_S$, and $G_0 R_D$, where $G_0 \equiv 2qZ\mu_n N_D a/L$.) Run your program assuming $V_{bi} = 1$ V and $V_P = -2.5$ V; successively set $G_0 R_S = G_0 R_D = 0$, 0.1, and 0.5. Compare your results with Fig. 1.10. Comment on the comparison.

2 / MOS Fundamentals

The metal–oxide (SiO_2)–semiconductor (Si) or MOS structure is, without a doubt, the core structure in modern-day microelectronics. Even ostensibly *pn* junction type devices incorporate the MOS structure in some functional and/or physical manner. A quasi-MOS device, as noted in the General Introduction, was first proposed in the 1920s. The dawn of modern history, however, is generally attributed to D. Kahng and M. M. Atalla who filed for patents on the Si–SiO_2 based field-effect transistor in 1960. The MOS designation, it should be noted, is reserved for the technologically dominant metal–SiO_2–Si system. The more general designation, metal–insulator–semiconductor (MIS), is used to identify similar device structures composed of an insulator other than SiO_2 or a semiconductor other than Si.

This chapter is intended to serve as an introduction to MOS structural and device fundamentals. The two-terminal MOS-capacitor or MOS-C is both the simplest of MOS devices and the structural heart of all MOS devices. We begin with a precise specification of the "ideal" MOS-C structure. Energy band and block-charge diagrams are next constructed and utilized to qualitatively visualize the charge, electric field, and band bending inside the MOS-C under static biasing conditions. Quantitative relationships for the electrostatic variables inside the semiconductor are then developed and subsequently related to the voltage applied to the metallic gate. Capacitance is of course the primary electrical observable exhibited by an MOS-capacitor. The MOS-C capacitance–voltage ($C–V$) characteristics are important not only from a fundamental but also a practical viewpoint. In the final section of the chapter, our knowledge of the internal workings of the MOS structure is used to explain and analyze the normally observed form of the MOS-C $C–V$ characteristics. The chapter concludes with an examination of computed ideal-structure characteristics, comments about measurement procedures, and other relevant $C–V$ considerations.

2.1 IDEAL STRUCTURE DEFINITION

As pictured in Fig. 2.1, the MOS-capacitor is a simple two-terminal device composed of a thin (0.01 μm–1.0 μm) SiO_2 layer sandwiched between a silicon substrate and a metallic field plate. The most common field plate materials are aluminum and heavily

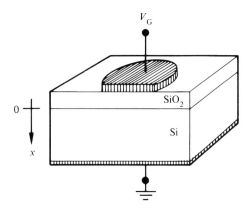

Fig. 2.1   The metal–oxide–semiconductor capacitor.

doped polycrystalline silicon.* A second metallic layer present along the back or bottom side of the semiconductor provides an electrical contact to the silicon substrate. The terminal connected to the field plate and the field plate itself are referred to as the gate; the silicon-side terminal, which is normally grounded, is simply called the back or substrate contact.

The ideal MOS structure has the following explicit properties: (1) the metallic gate is sufficiently thick so that it can be considered an equipotential region under ac as well as dc biasing conditions; (2) the oxide is a *perfect insulator* with *zero current* flowing through the oxide layer under *all* static biasing conditions; (3) there are no charge centers located in the oxide or at the oxide–semiconductor interface; (4) the semiconductor is uniformly doped; (5) the semiconductor is sufficiently thick so that, regardless of the applied gate potential, a field-free region (the so-called Si "bulk") is encountered before reaching the back contact; (6) an *ohmic* contact has been established between the semiconductor and the metal on the back side of the device; (7) the MOS-C is a one-dimensional structure with all variables taken to be a function only of the x-coordinate (see Fig. 2.1); and (8) $\Phi_M = \chi + (E_c - E_F)_\infty$, where Φ_M, χ, and $(E_c - E_F)_\infty$ are material parameters (energies) defined in Fig. 2.2. (A discussion of these material parameters is presented in the next section.)

All of the listed idealizations can be approached in practice and the ideal MOS structure is fairly realistic. For example, the resistivity of SiO_2 can be as high as 10^{18} ohm-cm, and the dc leakage current through the layer is indeed negligible for typical oxide thicknesses and applied voltages. Moreover, even very thin gates can be considered equipotential regions and ohmic back contacts are quite easy to achieve in practice. Similar statements can be made concerning most of the other idealizations.

---

*Heavily doped Si is metallic in nature. Poly-Si gates, used extensively in complex MOS device structures, are deposited by a chemical-vapor process and are then heavily doped by either phosphorus or boron diffusion.

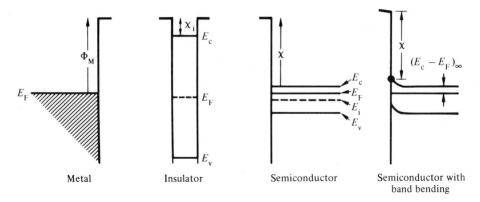

Fig. 2.2 Individual energy band diagrams for the metal, insulator, and semiconductor components of the MOS structure. The diagram labeled "semiconductor with band bending" defines $(E_c - E_F)_\infty$ and shows χ to be invariant with band bending. The value of χ, it should be emphasized, is measured relative to E_c at the semiconductor surface.

Special note, however, should be made of idealization 8. The $\Phi_M = \chi + (E_c - E_F)_\infty$ requirement could be omitted and will in fact be eliminated in Chapter 4. The requirement has only been included at this point to avoid unnecessary complications in the initial description of the static behavior.

2.2 ELECTROSTATICS—MOSTLY QUALITATIVE

2.2.1 Visualization Aids

Energy Band Diagram

The energy band diagram is an indispensable aid in visualizing the internal status of the MOS structure under static biasing conditions. The task at hand is to construct the diagram appropriate for the ideal MOS structure under equilibrium (zero-bias) conditions.

Figure 2.2 shows the surface-included energy band diagrams for the individual components of the MOS structure. In each case the abrupt termination of the diagram in a vertical line is meant to imply a surface. The ledge at the top of the vertical line, known as the vacuum level, denotes the minimum energy an electron must possess to completely free itself from the material. The energy difference between the vacuum level and the Fermi energy in a metal is known as the metal workfunction, Φ_M. In the semiconductor the height of the surface energy barrier is specified in terms of the electron affinity, χ, the energy difference between the vacuum level and the conduction band edge at the surface. χ is used instead of $E_{vacuum} - E_F$ because the latter quantity is not a constant in semiconductors, but varies as a function of doping and band bending near the surface. The remaining component, the insulator, is in essence modeled as an intrinsic wide-gap semiconductor where the surface barrier is again specified in terms of the electron affinity.

The conceptual formation of the MOS zero-bias band diagram from the individual components involves a two-step process. First the metal and semiconductor are brought together until they are a distance x_o apart. The insulator of thickness x_o is then inserted into the empty space between the metal and semiconductor components. Since the Fermi level must line up in any structure under equilibrium conditions (see Subsection 3.2.4, Volume I), and because we have specified $\Phi_M = \chi + (E_c - E_F)_\infty$, the vacuum levels of the M and S components are always in perfect alignment. The foregoing implies that there are no charges or electric fields induced in the system as the metal and semiconductor are brought together. Moreover, given the zero electric field in the x_o gap, the only effect of inserting the insulator is to slightly lower the barrier between the M and S components. Thus the equilibrium energy band diagram for the ideal MOS structure is concluded to be of the form pictured in Fig. 2.3.

Block Charge Diagrams

Complementary in nature to the energy band diagram, block charge diagrams provide information about the approximate charge distribution inside the MOS structure. As just noted in the energy band diagram discussion, there are no charges anywhere inside the ideal MOS structure under equilibrium conditions. However, when a bias is applied to the MOS-C, charge appears within the metal and semiconductor near the metal–oxide and oxide–semiconductor interfaces. A sample block charge diagram is shown in Fig. 2.4. Note that no attempt is made to represent the exact charge distributions inside the structure. Rather, a squared-off or block approximation is employed and hence the resulting figure is called a block charge diagram. Block charge diagrams are intended to be qualitative in nature; the magnitude and spatial extent of the charges should be interpreted with this fact in mind. Nevertheless, because the electric field is zero in the interior of both the metal and the semiconductor (see idealization 5), the charges within the structure must sum to zero according to Gauss's law. Consequently, in con-

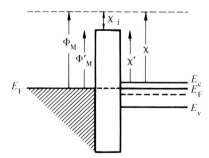

Fig. 2.3   Equilibrium energy band diagram for an ideal MOS structure.

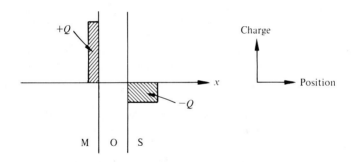

Fig. 2.4   Sample block charge diagram.

structing block charge diagrams the area representing positive charges is always drawn equal to the area representing negative charges.

2.2.2   Effect of an Applied Bias

General Observations

Before examining specific-case situations, it is useful to establish general ground rules as to how one modifies the MOS energy band diagram in response to an applied bias. Assume normal operating conditions where the back side of the MOS-C is grounded and let V_G be the dc bias applied to the gate.

With $V_G \neq 0$ we note first of all that the *semiconductor Fermi energy is unaffected by the bias and remains invariant (level on the diagram) as a function of position.* This is a direct consequence of the assumed zero current flow through the structure under all static biasing conditions. In essence, the semiconductor always remains in equilibrium independent of the bias applied to the MOS-C gate. Second, as in a *pn* junction, the applied bias separates the Fermi energies at the two ends of the structure by an amount equal to qV_G; that is,

$$E_F(\text{metal}) - E_F(\text{semiconductor}) = -qV_G \qquad (2.1)$$

Conceptually, the metal and semiconductor Fermi levels may be thought of as "handles" connected to the outside world. In applying a bias, one grabs onto the handles and re-arranges the relative up-and-down positioning of the Fermi levels. The back contact is grounded and the semiconductor-side handle therefore remains fixed in position. The metal-side handle, on the other hand, is moved downward if $V_G > 0$ and upward if $V_G < 0$.

Since the barrier heights are fixed quantities, the movement of the metal Fermi level obviously leads in turn to a distortion in other features of the band diagram. The situation is akin to bending a rubber doll out of shape. Viewed another way, $V_G \neq 0$ causes potential drops and E_c (E_v) band bending interior to the structure. No band bending occurs, of course, in the metal because it is an equipotential region. In the oxide and semiconductor, however, the energy bands must exhibit an upward slope (increasing E going from the gate toward the back contact) when $V_G > 0$ and a downward slope when $V_G < 0$. Moreover, the application of Poisson's equation to the oxide, taken to be an ideal insulator with no carriers or charge centers, yields $d\mathscr{E}_{oxide}/dx = 0$ and therefore $\mathscr{E}_{oxide} = $ constant. Hence, the slope of the energy bands in the oxide is a constant—E_c and E_v are linear functions of position. Naturally, band bending in the semiconductor is expected to be somewhat more complex in its functional form, but per idealization 5, must always vanish ($\mathscr{E} \rightarrow 0$) before reaching the back contact.

Specific Biasing Regions

Given the general principles just discussed, it is now a relatively simple matter to describe the internal status of the ideal MOS structure under various static biasing conditions. Taking the Si substrate to be n-type, consider first the application of a positive bias. The application of $V_G > 0$ lowers E_F in the metal relative to E_F in the semiconductor and causes a positive sloping of the energy bands in both the insulator and semiconductor. The resulting energy band diagram is shown in Fig. 2.5(a). The major conclusion to be derived from Fig. 2.5(a) is that the electron concentration inside the semiconductor, $n = n_i \exp[(E_F - E_i)/kT]$, increases as one approaches the oxide–semiconductor interface. This particular situation, where the majority carrier concentration is greater near the oxide–semiconductor interface than in the bulk of the semiconductor, is known as *accumulation*.

When viewed from a charge standpoint, the application of $V_G > 0$ places positive charges on the MOS-C gate. To maintain a balance of charge, negatively charged electrons must be drawn toward the semiconductor–insulator interface — the same conclusion established previously by using the energy band diagram. Thus the charge inside the device as a function of position can be approximated as shown in Fig. 2.5(b).

Consider next the application of a *small* negative potential to the MOS-C gate. The application of a small $V_G < 0$ slightly raises E_F in the metal relative to E_F in the semiconductor and causes a small negative sloping of the energy bands in both the insulator and semiconductor, as displayed in Fig. 2.5(c). From the diagram it is clear that the concentration of majority carrier electrons has been decreased, depleted, in the vicinity of the oxide–semiconductor interface. A similar conclusion results from charge considerations. Setting $V_G < 0$ places a minus charge on the gate, which in turn repels electrons from the oxide–semiconductor interface and exposes the positively charged donor sites. The approximate charge distribution is therefore as shown in Fig. 2.5(d). This situation, where the electron and hole concentrations at the oxide–semiconductor interface are less than the background doping concentration (N_A or N_D) is known for obvious reasons as *depletion*.

Finally, suppose a larger and larger negative bias is applied to the MOS-C gate. As V_G is increased negatively from the situation pictured in Fig. 2.5(c), the bands at the semi-

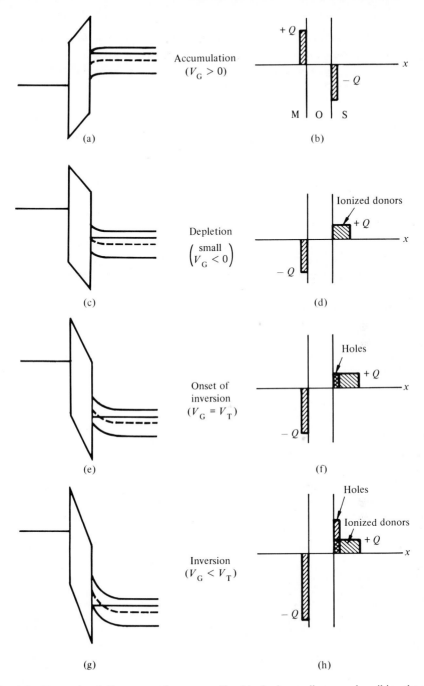

Fig. 2.5 Energy band diagrams and corresponding block charge diagrams describing the static state in an ideal *n*-type MOS-capacitor.

conductor surface will bend up more and the hole concentration at the surface (p_s) will likewise increase systematically from less than n_i when E_i (surface) $< E_F$, to n_i when E_i (surface) $= E_F$, to greater than n_i when E_i (surface) exceeds E_F. Eventually, the hole concentration increases to the point shown in Fig. 2.5(e) and (f), where

$$E_i(\text{surface}) - E_i(\text{bulk}) = 2[E_F - E_i(\text{bulk})] \qquad (2.2)$$

and

$$p_s = n_i e^{[E_i(\text{surface}) - E_F]/kT} = n_i e^{[E_F - E_i(\text{bulk})]/kT} = n_{\text{bulk}} = N_D \qquad (2.3)$$

Clearly, when $p_s = N_D$ for the special applied bias $V_G = V_T$ the surface is no longer depleted. Moreover, for further increases in negative bias ($V_G < V_T$), p_s exceeds $n_{\text{bulk}} = N_D$ and the surface region appears to change in character from n-type to p-type. In accordance with the change in character observation, the $V_G < V_T$ situation where the minority carrier concentration at the surface exceeds the bulk majority carrier concentration is referred to as *inversion*. Energy band and block charge diagrams depicting the inversion condition are displayed in Fig. 2.5(g) and (h).

If analogous biasing considerations are performed for an ideal p-type device, the results will be shown in Fig. 2.6. It is important to note from this figure that biasing regions

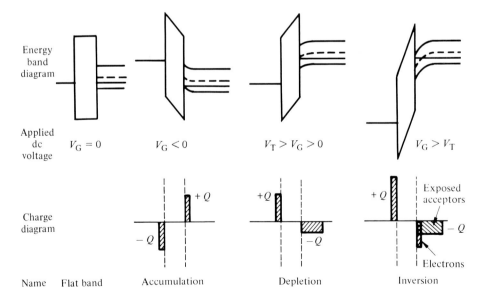

Fig. 2.6 Energy band and block charge diagrams for a p-type device under flat band, accumulation, depletion, and inversion conditions.

in a p-type device are reversed in polarity relative to the voltage regions in an n-type device; that is, accumulation in a p-type device occurs when $V_G < 0$, and so forth.

In summary, then, one can distinguish three physically distinct biasing regions— accumulation, depletion, and inversion. For an ideal n-type device, accumulation occurs when $V_G > 0$, depletion when $V_T < V_G < 0$, and inversion when $V_G < V_T$. The cited voltage polarities are simply reversed for an ideal p-type device. No band bending in the semiconductor or *flat band* at $V_G = 0$ marks the dividing line between accumulation and depletion. The dividing line at $V_G = V_T$ is simply called the depletion–inversion transition point, with Eq. (2.2) quantitatively specifying the onset of inversion for both n- and p-type devices.

2.3   ELECTROSTATICS—QUANTITATIVE FORMULATION

2.3.1   Semiconductor Electrostatics

Preparatory Considerations

The purpose of this section is to establish analytical relationships for the charge density (ρ), the electric field ($\mathscr{E}$), and the electrostatic potential existing inside an ideal MOS-C under static biasing conditions. The task is simplified by noting that the metal is an equipotential region. Charge appearing near the metal–oxide interface resides only a few Ångstroms (1 Ångstrom $= 10^{-8}$ cm) into the metal and, to a high degree of precision, may be modeled as a δ-function of charge at the M–O interface. Since by assumption there is no charge in the oxide (idealization 3), the magnitude of the charge in the metal is simply equal to the sum of the charges inside the semiconductor. Also, as noted previously, with no charges in the oxide, it follows that the electric field is constant in the oxide and the potential is a linear function of position. In other words, solving for the electrostatic variables inside an ideal MOS-C essentially reduces to solving for the electrostatic variables inside the semiconductor component of the MOS-C.

The mathematical description of the electrostatics inside the semiconductor is established in a relatively straightforward manner beginning with Poisson's equation. In the following analysis we invoke the depletion approximation to obtain a first-order closed-form solution. The development closely parallels the presentation in the *pn* junction analysis of Chapter 2, Volume II. It should be pointed out, however, that an *exact* solution is possible in the MOS-C case. The exact solution stems from simplifications associated with the fact that the semiconductor in an ideal MOS-C is *always* in equilibrium regardless of the applied dc bias. For reference purposes the exact solution is presented in Appendix B.

In performing the analysis, $\phi(x)$ is taken to be the potential inside the semiconductor at a given point x; x is understood to be the depth into the semiconductor as measured from the oxide–semiconductor interface [see Fig. 2.7(a)]. The symbol ϕ, instead of V, is used in MOS-C work to avoid possible confusion with externally applied potentials. In accordance with idealization 5, the electric field ($\mathscr{E} = -d\phi/dx$) is assumed to vanish as one proceeds into the semiconductor substrate. Following standard con-

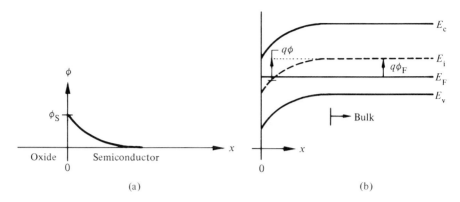

Fig. 2.7 Electrostatic parameters: (a) Graphical definition of ϕ and ϕ_S. (b) Relationship between $\phi(x)$ and band bending; graphical definition of ϕ_F.

vention, the potential is chosen to be zero in the field-free region of the substrate referred to as the semiconductor bulk. ϕ evaluated at the oxide–semiconductor interface (at $x = 0$) is given the special symbol, ϕ_S, and is known as the surface potential.

Figure 2.7(b) indicates how $\phi(x)$ is related to band bending on the energy band diagram. As shown (and consistent with Eq. (3.12) in Volume I),

$$\phi(x) = \frac{1}{q}[E_i(\text{bulk}) - E_i(x)] \tag{2.4}$$

and

$$\phi_S = \frac{1}{q}[E_i(\text{bulk}) - E_i(\text{surface})] \tag{2.5}$$

Fig. 2.7(b) also introduces an important material parameter; namely,

$$\phi_F \equiv \frac{1}{q}[E_i(\text{bulk}) - E_F] \tag{2.6}$$

ϕ_F is clearly related to the semiconductor doping. For one, the sign of ϕ_F indicates the doping type; directly from the definition one concludes $\phi_F > 0$ if the semiconductor is

p-type and $\phi_F < 0$ if the semiconductor is n-type. More importantly, the magnitude of ϕ_F is functionally related to the doping concentration. Given a nondegenerate Si substrate maintained at or near room temperature, we know from Volume I that

$$p_{\text{bulk}} = n_i e^{[E_i(\text{bulk}) - E_F]/kT} = N_A \qquad \ldots \text{if } N_A \gg N_D \qquad (2.7a)$$

$$n_{\text{bulk}} = n_i e^{[E_F - E_i(\text{bulk})]/kT} = N_D \qquad \ldots \text{if } N_D \gg N_A \qquad (2.7b)$$

Thus, combining Eqs. (2.6) and (2.7) yields

$$\phi_F = \begin{cases} \dfrac{kT}{q} \ln(N_A/n_i) & \ldots p\text{-type semiconductor} & (2.8a) \\[2ex] -\dfrac{kT}{q} \ln(N_D/n_i) & \ldots n\text{-type semiconductor} & (2.8b) \end{cases}$$

Extensive use will be made of the ϕ_S and ϕ_F parameters throughout the MOS discussion. Our immediate interest in these parameters involves their use in quantitatively specifying the biasing state inside the semiconductor. Clearly, under flat band conditions $\phi_S = 0$. Moreover, substituting Eqs. (2.5) and (2.6) into Eq. (2.2), one concludes

$$\phi_S = 2\phi_F \qquad \text{at the depletion–inversion transition point} \qquad (2.9)$$

With $\phi_F > 0$ in a p-type semiconductor, it follows that $\phi_S < 0$ if the semiconductor is accumulated, $0 < \phi_S < 2\phi_F$ if the semiconductor is depleted, and $\phi_S > 2\phi_F$ if the semiconductor is inverted. For an n-type semiconductor the inequalities are merely reversed.

SEE EXERCISE 2.1 — APPENDIX A

Delta-Depletion Solution

The approximate closed-form solution for the electrostatic variables based in part on the depletion approximation is conveniently divided into three segments corresponding to the three biasing regions of accumulation, depletion, and inversion.

Let us first consider accumulation. Figure 2.8 displays charge density and potential plots constructed using the exact solution found in Appendix B. After verifying the general correlation between the Fig. 2.8 plots and the p-bulk semiconductor portion of

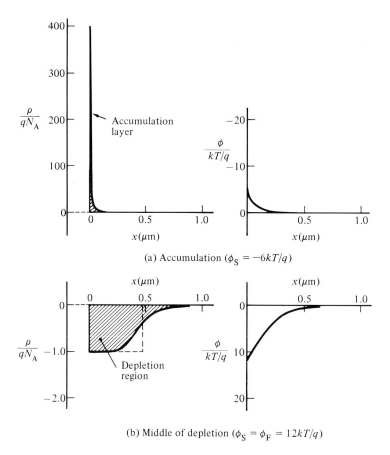

(a) Accumulation ($\phi_S = -6kT/q$)

(b) Middle of depletion ($\phi_S = \phi_F = 12kT/q$)

Fig. 2.8 Exact solution for the charge density and potential inside the semiconductor component of an MOS-C assuming $\phi_F = 12kT/q$ and $T = 300$ K ($kT/q = 0.259$ V). (a) Accumulation ($\phi_S = -6kT/q$), (b) middle of depletion ($\phi_S = \phi_F = 12kT/q$), (c) onset of inversion ($\phi_S = 2\phi_F = 24kT/q$), and (d) heavily inverted ($\phi_S = 2\phi_F + 6kT/q = 30kT/q$). The ρ-diagrams were drawn on a linear scale and the $+\phi$ axes oriented downward to enhance the correlation with the diagrams sketched in Fig. 2.6. The dashed-lines on the part (b) through (d) ρ-plots outline the depletion approximation version of the charge distribution.

the diagrams sketched in Fig. 2.6, specifically note from Fig. 2.8(a) that *the charge associated with majority carrier accumulation resides in an extremely narrow portion of the semiconductor immediately adjacent to the oxide–semiconductor interface.* By comparison, the depleted portion of the semiconductor under moderate depletion biasing shown in Fig. 2.8(b) extends much deeper into the semiconductor. Given the narrow extent of the accumulation layer, it would appear reasonable as a first order approximation to replace the accumulation charge with a δ-function of equal charge posi-

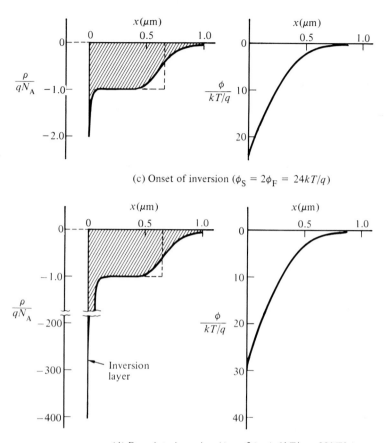

(c) Onset of inversion ($\phi_S = 2\phi_F = 24kT/q$)

(d) Deep into inversion ($\phi_S = 2\phi_F + 6kT/q = 30kT/q$)

Fig. 2.8   continued

tioned at the oxide–semiconductor interface. Indeed, we have just described the delta-depletion solution for accumulation. Because of the assumed δ-function of charge at $x = 0$, it automatically follows that the electric field and electrostatic potential are identically zero for all $x > 0$ under accumulation biasing in the delta-depletion solution. This is clearly somewhat inaccurate, but acceptable as a first-order approximation.

Turning next to inversion, we note from Fig. 2.8(d) that, like the accumulation layer charge, *the charge associated with minority carrier inversion resides in an extremely narrow portion of the semiconductor immediately adjacent to the oxide–semiconductor interface.* Moreover, in comparing the depleted semiconductor regions when $\phi_S = \phi_F$ (middle of depletion), $\phi_S = 2\phi_F$ (onset of inversion), and $\phi_S = 2\phi_F + 6kT/q$ (inversion), we find the depletion width increases substantially with increased depletion biasing, *but increases only slightly once the semiconductor inverts.* Based on the first of the foregoing observations, the actual inversion layer charge is approximately modeled in the delta-depletion solution by a δ-function of equal charge positioned at the oxide–semiconductor interface. To account for the second observation, it is additionally assumed the δ-function of charge added in inversion *precisely* balances the charge added to the MOS-C gate. As a consequence, in the delta-depletion solution for inverstion biases, the depletion region charge, the $x > 0$ electric field, and the $x > 0$ electrostatic potential remain fixed at their $\phi_S = 2\phi_F$ values. In other words, the inversion bias solution is established by merely adding a δ-function of surface charge to the solution existing at the end of depletion.

The remaining biasing region to be considered is depletion. In the standard depletion approximation the actual depletion charge is replaced with a squared-off distribution terminated abruptly a distance $x = W$ into the semiconductor. Assuming a p-type semiconductor and invoking the depletion approximation, one can write

$$\rho = q(p - n + N_D - N_A) \cong -qN_A \qquad (0 \leq x \leq W) \tag{2.10}$$

Poisson's equation then reduces to

$$\frac{d\mathscr{E}}{dx} = \frac{\rho}{K_S \varepsilon_0} \cong -\frac{qN_A}{K_S \varepsilon_0} \qquad (0 \leq x \leq W) \tag{2.11}$$

The straightforward integration of Eq. (2.11) employing the boundary condition $\mathscr{E} = 0$ at $x = W$ next yields

$$\mathscr{E}(x) = -\frac{d\phi}{dx} = \frac{qN_A}{K_S \varepsilon_0}(W - x) \qquad (0 \leq x \leq W) \tag{2.12}$$

A second integration with $\phi = 0$ at $x = W$ gives

$$\phi(x) = \frac{qN_A}{2K_S \varepsilon_0}(W - x)^2 \qquad (0 \leq x \leq W) \tag{2.13}$$

The final unknown in the electrostatic relationships, the depletion width W, is determined from Eq. (2.13) by applying the boundary condition $\phi = \phi_S$ at $x = 0$. We obtain

$$\phi_S = \frac{qN_A}{2K_S \varepsilon_0}W^2 \tag{2.14}$$

and therefore

$$W = \left[\frac{2K_S \varepsilon_0}{qN_A}\phi_S\right]^{1/2} \tag{2.15}$$

Taken together, Eqs. (2.10), (2.12), (2.13), and (2.15) constitute the desired depletion bias solution. For an n-bulk device N_A in the preceding equations is replaced by $-N_D$.

Before concluding, special note should be made of the depletion width, W_T, existing at the depletion–inversion transition point. In the delta-depletion formulation W_T is of course the maximum attainable equilibrium depletion width. Since $W = W_T$ when $\phi_S = 2\phi_F$, simple substitution into Eq. (2.15) yields

$$W_T = \left[\frac{2K_S \varepsilon_0}{qN_A} (2\phi_F) \right]^{1/2} \qquad (2.16)$$

A plot of W_T versus doping covering the typical range of MOS doping concentrations is displayed in Fig. 2.9.

2.3.2 Gate Voltage Relationship

In subsection 2.3.1 the biasing state was described in terms of the semiconductor surface potential, ϕ_S. Results formulated in this manner are dependent only on the properties of the semiconductor. ϕ_S, however, is an *internal* system constraint or boundary

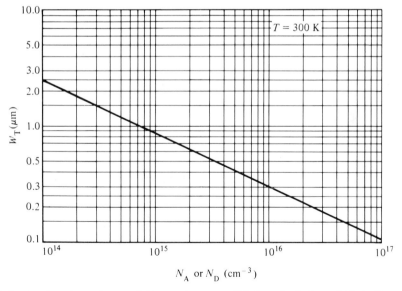

Fig. 2.9 Doping dependence of the maximum equilibrium depletion width inside silicon devices maintained at 300 K.

condition. It is the *externally* applied gate potential, V_G, which is subject to direct control. Thus, if the results of subsection 2.3.1 are to be utilized in practical problems, an expression relating V_G and ϕ_S must be established. This subsection is devoted to deriving the required relationship.

We begin by noting that V_G in the ideal structure is dropped partly across the oxide and partly across the semiconductor, or symbolically,

$$V_G = \Delta\phi_{semi} + \Delta\phi_{ox} \tag{2.17}$$

Because $\phi = 0$ in the semiconductor bulk, however, the voltage drop across the semiconductor is simply

$$\Delta\phi_{semi} = \phi(x = 0) = \phi_S \tag{2.18}$$

The task of developing a relationship between V_G and ϕ_S is therefore reduced to the problem of expressing $\Delta\phi_{ox}$ in terms of ϕ_S.

As stated previously (Subsection 2.2.2), in an ideal insulator with no carriers or charge centers

$$\frac{d\mathscr{E}_{ox}}{dx} = 0 \tag{2.19}$$

and

$$\mathscr{E}_{ox} = -\frac{d\phi_{ox}}{dx} = \text{constant} \tag{2.20}$$

Therefore

$$\Delta\phi_{ox} = \int_{-x_0}^{0} \mathscr{E}_{ox}\, dx = x_0 \mathscr{E}_{ox} \tag{2.21}$$

where x_0 is the oxide thickness. The next step is to relate $\mathscr{E}_{ox}$ to the electric field in the semiconductor. The well-known boundary condition on the fields normal to an interface between two dissimilar materials requires

$$(D_{semi} - D_{ox})|_{\text{O-S interface}} = Q_{O-S} \tag{2.22}$$

where $D = \varepsilon\mathscr{E}$ is the dielectric displacement and $Q_{O\text{-}S}$ is the charge/unit area located at the interface. Since $Q_{O\text{-}S} = 0$ in the idealized structure (idealization 3),*

$$D_{ox} = D_{semi}|_{x=0} \qquad (2.23)$$

$$\mathscr{E}_{ox} = \frac{K_S}{K_O}\mathscr{E}_S \qquad (2.24)$$

and

$$\Delta\phi_{ox} = \frac{K_S}{K_O}x_o\mathscr{E}_S \qquad (2.25)$$

K_S is the semiconductor dielectric constant; K_O, the oxide dielectric constant; and $\mathscr{E}_S$, the electric field in the semiconductor at the oxide–semiconductor interface. Finally, substituting Eqs. (2.18) and (2.25) into Eq. (2.17), and recognizing that $\mathscr{E}_S$ is a known or readily determined function of ϕ_S, we obtain

$$\boxed{V_G = \phi_S + \frac{K_S}{K_O}x_o\mathscr{E}_S} \qquad (2.26)$$

If the results of the delta-depletion solution are employed, a combination of Eqs. (2.12) and (2.15) gives

$$\mathscr{E}_S = \left[\frac{2qN_A}{K_S\varepsilon_0}\phi_S\right]^{1/2} \qquad (2.27)$$

and

$$V_G = \phi_S + \frac{K_S}{K_O}x_o\sqrt{\frac{2qN_A}{K_S\varepsilon_0}\phi_S} \qquad (0 \le \phi_S \le 2\phi_F) \qquad (2.28)$$

The $V_G - \phi_S$ dependence calculated from Eq. (2.28) employing a typical set of device parameters is displayed in Fig. 2.10. Also shown is the corresponding exact dependence. The figure nicely illustrates certain important features of the gate voltage rela-

---

*If the delta-depletion formulation is invoked, the δ-function layers of carrier charge at the O–S interface would technically contribute a $Q_{O\text{-}S}$ under accumulation and inversion conditions. However, $\phi_S = 0$ for all accumulation biases and $\phi_S = 2\phi_F$ for all inversion biases in the delta-depletion solution. In the cited formulation, therefore, the $V_G - \phi_S$ relationship we are deriving would only be used in performing depletion-bias calculations.

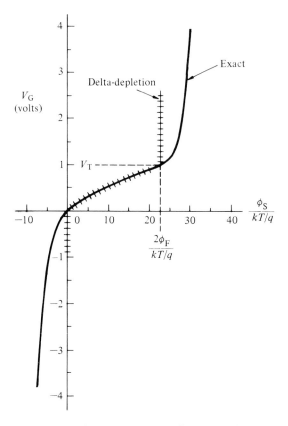

Fig. 2.10 Typical interrelationship between the applied gate voltage and the semiconductor surface potential; +++++++ delta-depletion solution, ——— exact solution ($x_o = 0.1\ \mu m$, $N_A = 10^{15}/cm^3$, $T = 300$ K).

tionship. For one, ϕ_S is a rather rapidly varying function of V_G when the device is depletion biased. However, when the semiconductor is accumulated ($\phi_S < 0$) or inverted ($\phi_S > 2\phi_F$), it takes a large change in gate voltage to produce a small change in ϕ_S. This implies the gate voltage divides proportionally between the oxide and the semiconductor under depletion biasing. Under accumulation and inversion biasing, on the other hand, changes in the applied potential are dropped almost totally across the oxide. Also note that the depletion bias region is only slightly greater than 1 volt in extent. Since the character of the semiconductor changes drastically in progressing from one side of the depletion bias region to the other, we are led to anticipate a significant variation in the electrical characteristics over a rather narrow range of voltages.

SEE EXERCISE 2.2 — APPENDIX A

2.4 CAPACITANCE — VOLTAGE CHARACTERISTICS

With dc current flow blocked by the oxide, the major observable exhibited by MOS-Cs is capacitance. As it turns out, the measured capacitance varies as a function of the applied gate voltage and the capacitance–voltage (C–V) characteristic is of considerable practical importance. To the device specialist, the MOS-C C–V characteristic is like a picture window, a window revealing the internal nature of the structure. The characteristic serves as a powerful diagnostic tool for identifying deviations from the ideal in both the oxide and the semiconductor. MOS-C C–V characteristics are routinely monitored during MOS device fabrication.

In most laboratories and fabrication facilities, the C–V measurements are performed with automated equipment. The device is positioned on a probing station (normally housed in a light-tight box to exclude room light) and is connected by shielded wires to a C–V bridge. The bridge superimposes a small ac signal on top of a preselected dc voltage and monitors the resulting ac current flowing into the test structure. The ac signal is typically less than 15 mV rms and a common signal frequency is 1 MHz. An analog output proportional to the capacitance determined by the bridge and the dc bias voltage are then fed into a calibrated X–Y recorder. Alternatively, digital C–V output is sent to a printer or plotter. Automatic provisions are made for slowly changing the dc voltage to obtain a continuous (or quasi-continuous) capacitance versus voltage characteristic.

This section is primarily devoted to modeling the observed form of the MOS-C C–V characteristic in the so-called low-frequency and high-frequency limits. These limiting-case designations refer to the frequency of the ac signal used in the capacitance measurement. The theoretical treatment is of course restricted in scope to the ideal structure. Practical measurement considerations, however, are included at the end of the section.

2.4.1 Theory and Analysis

Qualitative Theory

High- and low-frequency C–V data derived from a representative MOS-capacitor are displayed in Fig. 2.11. To explain the observed form of the C–V characteristics, let us consider how the charge inside an n-type MOS-C responds to the applied ac signal as the dc bias is systematically changed from accumulation, through depletion, to inversion. We begin with accumulation. In accumulation the dc state is characterized by the pile-up of majority carriers right at the oxide–semiconductor interface. Furthermore, under accumulation conditions the state of the system can be changed very rapidly. For typical semiconductor dopings, the majority carriers, the only carriers involved in the operation of the accumulated device, can equilibrate with a time constant on the order of 10^{-10} to 10^{-13} sec. Consequently, at standard probing frequencies of 1 MHz or less it is reasonable to assume the device can follow the applied ac signal quasi-statically, with the small ac signal adding or subtracting a small ΔQ on the two sides of the oxide as shown in Fig. 2.12(a). Since the ac signal merely adds or subtracts a charge close to

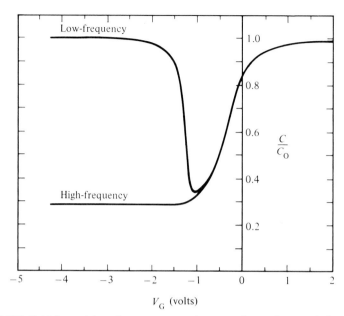

Fig. 2.11 MOS-C high- and low-frequency capacitance–voltage characteristics. The high-frequency curve was derived from a C–V bridge at a measurement frequency of 1 MHz. The low-frequency curve was obtained employing the slow-ramp or quasi-static technique. The device was fabricated on $N_D = 9.1 \times 10^{14}/\text{cm}^3$ (100) Si; $x_o = 0.119\mu m$.

the edges of an insulator, the charge configuration inside the accumulated MOS-C is essentially that of an ordinary parallel-plate capacitor. For either low or high probing frequencies we therefore conclude

$$C(\text{acc}) \simeq C_O = \frac{K_O \varepsilon_0 A_G}{x_o} \qquad (2.29)$$

where A_G is the area of the MOS-C gate.

Under depletion biasing the dc state of an n-type MOS structure is characterized by a $-Q$ charge on the gate and a $+Q$ depletion layer charge in the semiconductor. The depletion layer charge is directly related, of course, to the withdrawal of majority carriers from an effective width W adjacent to the oxide–semiconductor interface. Thus, once again, only majority carriers are involved in the operation of the device and the charge state inside the system can be changed very rapidly. As pictured in Fig. 2.12(b), when the ac signal places an increased negative charge on the MOS-C gate, the depletion layer inside the semiconductor widens almost instantaneously; that is, the depletion width quasi-statically fluctuates about its dc value in response to the applied ac signal.

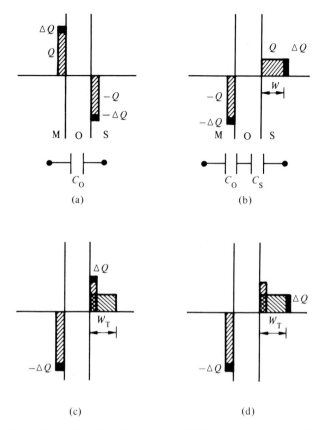

Fig. 2.12 ac charge fluctuations inside an *n*-type MOS-capacitor under dc biasing conditions corresponding to (a) accumulation, (b) depletion, (c) inversion when $\omega \to 0$, and (d) inversion when $\omega \to \infty$. Equivalent circuit models appropriate for accumulation and depletion biasing are also shown beneath the block charge diagrams in parts (a) and (b), respectively.

If the stationary dc charge in Fig. 2.12(b) is conceptually eliminated, all that remains is a small fluctuating charge on the two sides of a double-layer insulator. For all probing frequencies this situation is clearly analogous to two parallel plate capacitors (C_O and C_S) in series, where

$$C_O = \frac{K_O \varepsilon_0 A_G}{x_o} \quad \text{(oxide capacitance)} \tag{2.30a}$$

$$C_S = \frac{K_S \varepsilon_0 A_G}{W} \quad \text{(semiconductor capacitance)} \tag{2.30b}$$

and

$$C(\text{depl}) = \frac{C_0 C_S}{C_0 + C_S} = \frac{C_0}{1 + \dfrac{K_0 W}{K_S x_o}} \tag{2.31}$$

Note from Eq. (2.31) that, because W increases with increased depletion biasing, $C(\text{depl})$ correspondingly decreases as the dc bias is changed from flat band to the onset of inversion.

Once inversion is achieved we know that an appreciable number of minority carriers pile up near the oxide–semiconductor interface in response to the applied dc bias. Also, the dc width of the depletion layer tends to maximize at W_T. The ac charge response, however, is not immediately obvious. The inversion layer charge might conceivably fluctuate in response to the ac signal as illustrated in Fig. 2.12(c). Alternatively, the semiconductor charge required to balance ΔQ changes in the gate charge might result from small variations in the depletion width as pictured in Fig. 2.12(d). Even a combination of the two extremes is a logical possibility. The problem is to ascertain which alternative describes the actual ac charge fluctuation inside an MOS-C. As it turns out, the observed charge fluctuation depends on the frequency of the ac signal used in the capacitance measurement.

First of all, if the measurement frequency is very low ($\omega \to 0$), minority carriers can be generated or annihilated in response to the applied ac signal and the time-varying ac state is essentially a succession of dc states. Just as in accumulation, one has a situation (Fig. 2.12(c)) where charge is being added or subtracted close to the edges of a single-layer insulator. We therefore conclude

$$C(\text{inv}) \simeq C_0 \qquad \text{for } \omega \to 0 \tag{2.32}$$

If, on the other hand, the measurement frequency is very high ($\omega \to \infty$), the relatively sluggish generation–recombination process will not be able to supply or eliminate minority carriers in response to the applied ac signal. The number of minority carriers in the inversion layer therefore remains fixed at its dc value and the depletion width simply fluctuates about the W_T dc value. Similar to depletion biasing, this situation (Fig. 2.12(d)) is equivalent to two parallel-plate capacitors in series and

$$C(\text{inv}) = \frac{C_0 C_S}{C_0 + C_S} = \frac{C_0}{1 + \dfrac{K_0 W_T}{K_S x_o}} \qquad \text{for } \omega \to \infty \tag{2.33}$$

W_T is of course a constant independent of the dc inversion bias and $C(\text{inv})_{\omega \to \infty} = C(\text{depl})_{\text{minimum}} = $ constant for all inversion biases. Finally, if the measurement frequency is such that a *portion* of the inversion layer charge can be created/annihilated in response to the ac signal, an inversion capacitance intermediate between the high- and low-frequency limits will be observed.

An overall theory can now be constructed by combining the results of the foregoing accumulation, depletion, and inversion considerations. Specifically, we expect the MOS-C capacitance to be approximately constant at C_O under accumulation biases, to decrease as the dc bias progresses through depletion, and to be approximately constant again under inversion biases at a value equal to $\sim C_O$ if $\omega \to 0$ or $C(\text{depl})_{\min}$ if $\omega \to \infty$. Moreover, for an n-type device, accumulating gate voltages (where $C \simeq C_O$) are positive, inverting gate voltages are negative, and the decreasing-capacitance, depletion bias region is on the order of a volt or so in width. Quite obviously, this theory for the capacitance–voltage characteristics is in good agreement with the experimental MOS-C C–V_G characteristics presented in Fig. 2.11.

SEE EXERCISE 2.3 — APPENDIX A

Delta-Depletion Analysis

Building on the development in the previous subsection, it is relatively easy to establish a first-order quantiative theory based on the delta-depletion formulation. Specifically, in the delta-depletion formulation the charge blocks representing accumulation and inversion layers in Fig. 2.12 are formally replaced by δ-functions of charge *right at* the oxide–semiconductor interface. Consequently, C in the delta-depletion solution is *precisely* equal to C_O for accumulation biases and for inversion biases in the low-frequency limit. On the other hand, the depletion relationship and the high-frequency inversion relationship [Eqs. (2.31) and (2.33), respectively] can be used without modification. The block charge modeling of the depletion regions in Fig. 2.12 conforms exactly with the simplified charge distributions assumed in the depletion approximation. Within the framework of the delta-depletion formulation, therefore

$$
C = \begin{cases}
C_O & \text{acc} & (2.34a) \\[2ex]
\dfrac{C_O}{1 + \dfrac{K_O W}{K_S x_o}} & \text{depl} & (2.34b) \\[3ex]
C_O & \text{inv } (\omega \to 0) & (2.34c) \\[2ex]
\dfrac{C_O}{1 + \dfrac{K_O W_T}{K_S x_o}} & \text{inv } (\omega \to \infty) & (2.34d)
\end{cases}
$$

Given a set of device parameters, one can compute C_O and W_T from previous relationships. For the analytical solution to be complete, however, the depletion-bias W in Eq. (2.34b) must be expressed as a function of V_G. Inverting Eq. (2.28) to obtain ϕ_S (or more precisely, $\sqrt{\phi_S}$) as a function of V_G, and then substituting the result into Eq. (2.15), yields the required expression. We find

$$W = \frac{K_S}{K_O} x_o \left[\sqrt{1 + \frac{V_G}{V_\delta}} - 1 \right] \qquad (2.35)$$

where

$$V_\delta \equiv \frac{q}{2} \frac{K_S x_o^2}{K_O^2 \varepsilon_0} N_A \qquad \begin{array}{l} \ldots p\text{-bulk device} \\ (\text{for } n\text{-bulk } N_A \rightarrow -N_D) \end{array} \qquad (2.36)$$

Note that if Eq. (2.35) is substituted into Eq. (2.34b) one obtains the very simple result

$$C = \frac{C_O}{\sqrt{1 + \dfrac{V_G}{V_\delta}}} \qquad \text{(depletion biases)} \qquad (2.37)$$

A sample set of low- and high-frequency $C-V$ characteristics constructed using the results of the delta-depletion analysis is displayed in Fig. 2.13.

2.4.2 Computations and Observations

Exact Computations

The delta-depletion characteristics, as typified by Fig. 2.13, are a rather crude representation of reality. The first-order theory does a credible job for gate voltages comfortably within a given biasing region, but fails badly in the neighborhood of the transition points going from accumulation to depletion and from depletion to inversion. A more accurate modeling of the observed characteristics is often required in practical applications and is established by working with the exact-charge distribution inside the MOS-capacitor. The results of the exact-charge analysis are presented in Appendix C. Although the derivation of the exact-charge relationships is beyond the scope of this text, the results themselves are quite tractable. Highly-accurate ideal-structure $C-V$ characteristics can be readily constructed.

A number of sample $C-V$ characteristics calculated using the exact-charge relationships are displayed in Figs. 2.14 to 2.16. These figures, respectively, exhibit the general effect of varying the doping concentration (Fig. 2.14), the oxide thickness (Fig. 2.15),

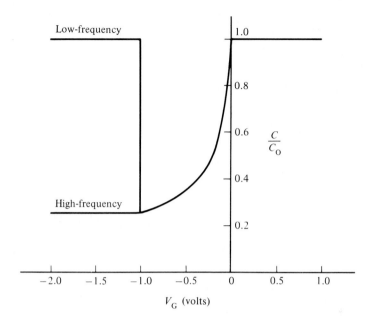

Fig. 2.13 Sample set of low- and high-frequency C–V characteristics established using the delta-depletion theory ($x_o = 0.1\ \mu$m, $N_D = 10^{15}/\text{cm}^3$, $T = 300$ K).

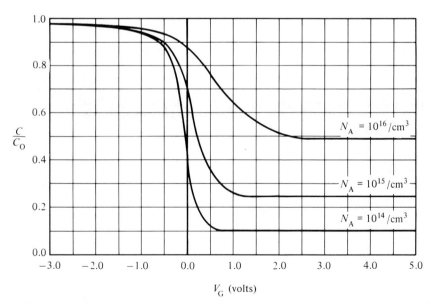

Fig. 2.14 Doping dependence of the high-frequency C–V_G characteristics. (Exact-charge theory $x_o = 0.1\ \mu$m, $T = 300$ K.)

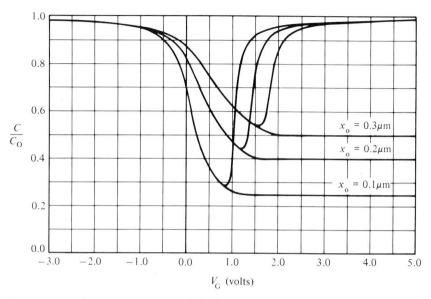

Fig. 2.15 Oxide thickness dependence of the low- and high-frequency C–V_G characteristics. (Exact-charge theory, $N_A = 10^{15}/cm^3$, $T = 300$ K.)

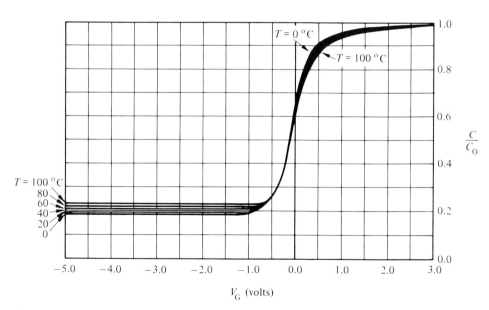

Fig. 2.16 Temperature dependence of the high-frequency C–V_G characteristics. (Exact-charge theory, $x_o = 0.1$ μm, $N_D = 5 \times 10^{14}/cm^3$.)

and the device temperature (Fig. 2.16). Note in particular from Fig. 2.14 the significant increase in the high-frequency inversion capacitance and the substantial widening of the depletion bias region with increased doping. In fact, at very high dopings (not shown) the capacitance approaches a constant independent of bias. This should not be an unexpected result, for with increased doping the semiconductor begins to look more and more like a metal and the MOS-C should be expected to react more and more like a standard capacitor. As illustrated in Fig. 2.15, an increase in the oxide thickness also widens the depletion bias region and affects the high-frequency inversion capacitance. The increased width of the depletion bias region with increased x_o is simply a consequence of a proportionate increase in the voltage drop across the oxide component of the structure. Finally, Fig. 2.16 nicely displays the moderate sensitivity of the inversion-bias capacitance and the near insensitivity of the depletion-bias capacitance to changes in temperature.

Practical Observations

In the discussion to this point we have more or less sidestepped any clarification of precisely what was meant by "low-frequency" and "high-frequency" in terms of actual measurement frequencies. One might wonder, will a 100 Hz ac signal typically yield low-frequency C–V characteristics? Perhaps surprisingly, the answer to the question is *no*. Given modern-day MOS-Cs with their long carrier lifetimes and low carrier generation rates, even probing frequencies as low as 10 Hz, the practical limit in bridge-type measurements, will yield high-frequency type characteristics. If an MOS-C low-frequency characteristic is required, indirect means such as the quasi-static technique* must be employed to construct the characteristic. In the quasi-static technique a slow (typically 10–100 mV/sec) linear voltage ramp is applied to the MOS-C gate and the *current* into the gate is monitored as a function of the gate voltage. As is readily confirmed, the quasi-static displacement current flowing through the device is directly proportional to the low-frequency capacitance; properly calibrated, the measured current versus voltage data replicates the desired low-frequency C–V characteristic.

On the high-frequency side, one cannot actually let $\omega \to \infty$ and expect to observe a high-frequency characteristic. Measurement frequencies, in fact, seldom exceed 1 MHz. At higher frequencies the resistance of the semiconductor bulk comes into play and lowers the observed capacitance. At even higher frequencies ($\gtrsim 1$ GHz) one must begin to worry about the response time of the majority carriers.

Normally, it is the high-frequency characteristic that is routinely recorded and the standard, almost universal measurement frequency is 1 MHz. This is not to say the high-frequency characteristics can be recorded at 1 MHz without exercising a certain amount of caution. Suppose, for example, the C–V measurement is performed as described earlier in this chapter, with the dc voltage being ramped from accumulation into inversion to obtain a continuous capacitance-versus-voltage characteristic. Figure 2.17

---

*See M. Kuhn, "A Quasi-Static Technique for MOS C–V and Surface State Measurements," *Solid-State Electronics*, **13**, 873 (1970).

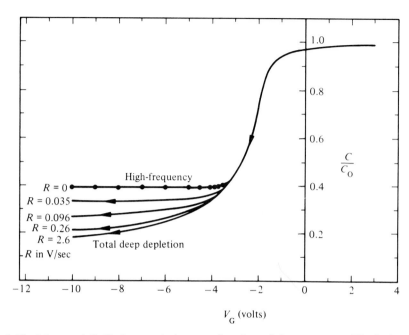

Fig. 2.17   Measured $C-V$ characteristics as a function of the ramp rate (R). In inversion the high-frequency capacitance was obtained by stopping the ramp and allowing the device to equilibrate.

illustrates the usual results of such a measurement performed at various ramp rates. Note that at even the slowest ramp rates one does not properly plot out the inversion portion of the high-frequency characteristic. One must stop the ramp in inversion and allow the device to equilibrate, or sweep the device backward from inversion toward accumulation, to accurately record the high-frequency inversion capacitance.

The discussion in the preceding paragraph really serves two purposes, the second being an entry into the important topic of deep depletion. Let us examine the ramped measurement (Fig. 2.17) in greater detail. When the ramp voltage is in accumulation or depletion, only majority carriers are involved in the operation of the device, and the dc charge configuration inside the structure rapidly reacts to the changing gate bias. As the ramp progresses from depletion into the inversion bias region, however, a significant number of minority carriers are required to attain an equilibrium charge distribution within the MOS-C. The minority carriers were not present prior to entering the inversion bias region, cannot enter the semiconductor from the remote back contact or across the oxide, and therefore must be created in the near-surface region of the semiconductor. The generation process, as we have noted several times, is rather sluggish and has difficulty supplying the minority carriers needed for the structure to equilibrate. Thus, as pictured in Fig. 2.18(a), the semiconductor is driven into a *nonequi-*

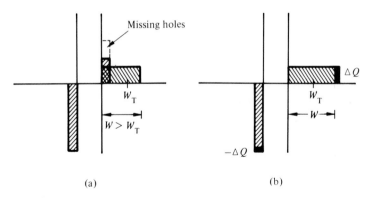

(a)   (b)

Fig. 2.18   (a) Nonequilibrium charge configuraton inside an n-type MOS-capacitor under deep-depletion conditions. (b) ac charge fluctuations inside the MOS-C when the semiconductor is totally deep depleted.

librium condition where, in balancing the charge added to the MOS-C gate, the depletion width becomes greater than W_T to offset the missing minority carriers. The described condition, the nonequilibrium condition where there is a deficit of minority carriers and a depletion width in excess of the equilibrium value, is referred to as *deep depletion*.

The existence of a $W > W_T$ of course explains the reduced values of C observed in the ramp measurement. Moreover, the decrease in capacitance with increased ramp rate indicates a greater deficit of minority carriers and a wider depletion width. This is logical, since the greater the ramp rate, the fewer the number of minority carriers generated prior to arriving at a given inversion bias.

The limiting case as far as deep-depletion is concerned occurs when the semiconductor is totally devoid of minority carriers — totally deep depleted. Except for a wider depletion width, the total deep depletion condition shown in Fig. 2.18(b) is precisely the same as the simple depletion condition pictured in Fig. 2.12(b). Consequently, by analogy, and based on the delta-depletion formulation, the limiting-case capacitance exhibited by the structure under deep depletion conditions should be

$$C = \frac{C_O}{\sqrt{1 + \dfrac{V_G}{V_\delta}}} \qquad \begin{matrix} \text{total deep depletion} \\ (V_G > V_T \; p\text{-type};\; V_G < V_T \; n\text{-type}) \end{matrix} \qquad (2.38)$$

Equation (2.38) is in excellent agreement with experimental observations and is essentially identical to the result obtained from an exact charge analysis. The 2.6 V/sec ramp rate curve shown in Fig. 2.17 is an example of a total-deep-depletion characteristic.

The deeply depleted condition, we should note, is central to the operation of the dynamic random access memory (DRAM) and the charge-coupled devices (CCDs) mentioned in the General Introduction. DRAMs use deeply depleted MOS-capacitors as storage elements. In CCD imagers, carrier charge is optically produced and temporarily stored in partially deep depleted potential wells beneath an array of MOS-C gates.

SEE EXERCISE 2.4 — APPENDIX A

2.5 SUMMARY AND CONCLUDING COMMENTS

This chapter was devoted to introducing basic MOS terminology, concepts, visualization aids, analytical procedures, and the like. We began by describing the MOS-capacitor and clearly defining what was envisioned as the ideal MOS structure. The ideal MOS-C serves not only as a convenient tool for the introduction of MOS fundamentals, but also provides a point of reference for understanding and analyzing the more complex behavior of real MOS structures. The reference nature of the ideal structure will become more apparent in Chapter 4 where some of the idealizations will be removed and the ensuring perturbations on the device characteristics will be carefully examined.

The internal state of the MOS-C under static-biasing conditions was qualitatively described using energy band and block charge diagrams. With the aid of the diagrams the terms accumulation, flat band, depletion, and inversion were given a physical interpretation. Accumulation corresponds to the pile-up of majority carriers at the oxide–semiconductor interface; flat band, to no bending of the semiconductor bands, or equivalently, to no charge in the semiconductor; depletion, to the repulsion to majority carriers from the interface leaving behind an uncompensated impurity-ion charge; and inversion, to the pile-up of minority carriers at the oxide–semiconductor interface.

As a point of information, it should be mentioned that, in certain analyses, it is convenient and reasonable to divide the depletion bias region as defined herein into two subregions. In some MOS publications one therefore finds the term "depletion" is only used in referring to band bendings between $\phi_S = 0$ and $\phi_S = \phi_F$. *Weak inversion* is used to describe band bendings from $\phi_S = \phi_F$ to $\phi_S = 2\phi_F$. In addition, *strong inversion* (implying more inversion than weak inversion) replaces inversion as defined herein.

Quantitative expression for the charge density, electric field, and electrostatic potential inside an MOS-C are established by solving Poisson's equation. A first-order solution based on the depletion approximation and δ-function modeling of the carrier charges was presented in the chapter proper. An exact solution for the electrostatic variables can be found in Appendix B.

The qualitative and quantitative formalism that had been developed was next applied to modeling the MOS-C C–V characteristics in the low- and high-frequency limits. Sample ideal-structure characteristics were presented and examined for parametric dependencies. Finally, the practical meaning of "low-frequency" and "high-frequency" was clarified, and deep depletion was noted to occur when the MOS-C is not allowed to equilibrate under inversion biasing. Deep depletion is a nonequilibrium condition where there is a deficit of minority carriers and a depletion width in excess of the equilibrium value.

PROBLEMS

2.1 For the ϕ_F, ϕ_S parameter sets listed below first indicate the specified biasing condition and then draw the energy band diagram and block charge diagram that characterize the static state of the system. Assume the MOS structure to be ideal.

(a) $\dfrac{\phi_F}{kT/q} = 12.5, \dfrac{\phi_S}{kT/q} = 30$

(b) $\dfrac{\phi_F}{kT/q} = -9, \dfrac{\phi_S}{kT/q} = 0$

(c) $\dfrac{\phi_F}{kT/q} = 12, \dfrac{\phi_S}{kT/q} = -2$

(d) $\dfrac{\phi_F}{kT/q} = 15, \dfrac{\phi_S}{kT/q} = 15$

(e) $\dfrac{\phi_F}{kT/q} = -10, \dfrac{\phi_S}{kT/q} = -20$

2.2 Let us examine Fig. 2.8, particularly Fig. 2.8(c), more closely.

(a) Draw the block charge diagram describing the charge situation inside an ideal p-bulk MOS-C biased at the onset of inversion.

(b) Is your part (a) diagram in agreement with the plot of ρ/qN_A versus x in Fig. 2.8(c)? Explain why the ρ/qN_A has a spikelike nature near $x = 0$ and shows a value of $\rho/qN_A = -2$ at $x = 0$.

(c) Noting that $\phi_F/(kT/q) = 12$ and $T = 300$ K was assumed in constructing Fig. 2.8, determine W_T. Is the deduced W_T consistent with the approximate charge distribution shown in Fig. 2.8(c)?

2.3 An MOS-C is maintained at $T = 300$ K ($K_S = 11.8$, $K_O = 3.9$, $kT/q = 0.0259$ V, $n_i = 1.18 \times 10^{10}/\mathrm{cm}^3$), $x_o = 0.1$ μm, and the Si doping is $N_D = 10^{15}/\mathrm{cm}^3$. Compute:

(a) ϕ_F (in kT/q units and in volts);

(b) W when $\phi_S = 2\phi_F$;

(c) $\mathscr{E}_S$ when $\phi_S = 2\phi_F$;

(d) V_G when $\phi_S = 2\phi_F$. (How is this result related to Fig. 2.10?)

2.4 (a) Making use of Appendix B, show that the exact-solution equivalent of Eq. (2.28) is

$$V_G = \frac{kT}{q}\left[U_S + \hat{U}_S \frac{K_S x_o}{K_O L_D} F(U_S, U_F)\right]$$

(b) Construct a computer program which calculates V_G as a function of U_S using the part (a) relationship. Only U_F and x_o are to be considered input variables; let $T = 300$ K and step U_S in one-unit increments from $U_S = U_F - 21$ to $U_S = U_F + 21$. Run the program assuming $x_o = 0.1$ μm and $N_D = 10^{15}/\text{cm}^3$. Compare your numerical results with the exact solution curve in Fig. 2.10.

2.5 With modern-day processing it is possible to produce semiconductor–oxide–semiconductor (SOS) capacitors in which a semiconductor replaces the metallic gate in a standard MOS-C. Answer the questions posed below assuming an SOS-C composed of two *identical p-type* nondegenerate silicon electrodes, an *ideal structure*, and a biasing arrangement as defined by Fig. P2.5. Include any comments which will help to forestall a misinterpretation of the requested pictorial answers.

(a) Draw the energy band diagrams for the structure when (i) $V_G = 0$, (ii) $V_G > 0$ but small, (iii) $V_G > 0$ and very large, (iv) $V_G < 0$ but small, and (v) $V_G < 0$ and very large.

(b) Draw the block charge diagrams corresponding to the five biasing conditions considered in part (a).

(c) Sketch the expected shape of the high-frequency $C-V_G$ characteristic for the SOS-C described in this problem. For reference purposes, also sketch on the same plot the high-frequency $C-V_G$ characteristic of an MOS-C assumed to have the same semiconductor doping and oxide thickness as the SOS-C.

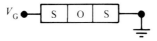

Fig. P2.5

2.6 (a) Following the approach suggested in the text, derive Eq. (2.35).

(b) Assuming $x_o = 0.1$ μm, $N_D = 10^{15}/\text{cm}^3$, and $T = 300$ K, compute:
   (i) W_T;
   (ii) C/C_O inversion ($\omega \to \infty$);
   (iii) V_T (delta-depletion theory).
   (iv) Comment on the comparison of your C and V results with Fig. 2.13.

2.7 The $C-V$ characteristic exhibited by an MOS-C (assumed to be ideal) is displayed in Fig. P2.7.

(a) Is the semiconductor component of the MOS-C doped n-type or p-type? Indicate how you arrived at your answer.

(b) Draw the MOS-C energy band diagram corresponding to point (2) on the $C-V$ characteristic. (Be sure to include the diagrams for all three components of the MOS-C, show the proper band bending in both the oxide and semiconductor, and properly position the Fermi level in the metal and semiconductor.)

(c) Draw the block charge diagram corresponding to point (1) on the $C-V$ characteristic.

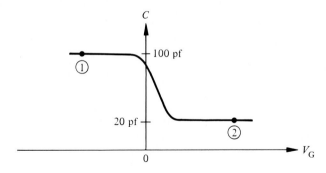

Fig. P2.7

(d) If $K_O = 3.9$ and the area of the MOS-C gate is 3×10^{-3} cm^2, what is the oxide thickness (x_o)?

(e) Determine W_T for the given MOS-C.

2.8 The dc state of an ideal MOS-capacitor is characterized by the block charge diagram shown in Fig. P2.8.

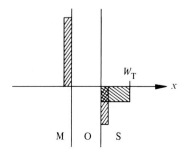

Fig. P2.8

(a) Is the semiconductor n- or p-type? Explain.

(b) Is the device accumulation, depletion, or inversion biased? Explain.

(c) Draw the energy band diagram corresponding to the charge state pictured in the block charge diagram.

(d) By appropriately modifying the block charge diagram, indicate how the charge state inside the MOS-C is modified when a high-frequency ac signal is applied to the device.

(e) Sketch the general shape of the high-frequency $C-V$ characteristic to be expected from the structure. Place an $\times$ on the $C-V$ characteristic at the point which corresponds to the charge state pictured in the Fig. P2.8 block charge diagram.

(f) While biased at the same gate voltage giving rise to the Fig. P2.8 diagram, the MOS-C is somehow *totally deep depleted.* Draw the block charge diagram describing the new state of the system.

2.9 (a) Employing the exact-charge relationships found in Appendix C, write a computer program that can be used to construct *low-frequency* C/C_O versus V_G characteristics. The program is to calculate C/C_O and the corresponding V_G for U_S stepped in one-unit increments from $U_S = U_F - 21$ to $U_S = U_F + 21$. Let $T = 300$ K, $kT/q = 0.0259$ V, $L_D = 2.68 \times 10^{-3}$ cm, $K_S = 11.8$, and $K_O = 3.9$. Only U_F (or N_A, N_D) and x_o are to be considered input variables.

(b) Setting $U_F = 11.35$ ($N_A = 10^{15}/\text{cm}^3$), use your part (a) program to compute C/C_O (low-frequency) versus V_G for $x_o = 0.1$ μm, 0.2 μm, and 0.3 μm. Compare your program results with Fig. 2.15.

(c) Employing the exact-charge relationships found in Appendix C, write a computer program that can be used to construct *high-frequency* C/C_O versus V_G characteristics. Follow the same general computational procedures noted in part (a). *Warning:* The high-frequency calculation is far more involved and computation intensive than the low-frequency calculation.

(d) Setting $x_o = 0.1$ μm, use your part (c) program to compute C/C_O (high frequency) versus V_G for $N_A = 10^{14}/\text{cm}^3$, $10^{15}/\text{cm}^3$, and $10^{16}/\text{cm}^3$. Compare your program results with Fig. 2.14.

2.10 The doping concentration (N_A or N_D) required in constructing the theoretical $C-V$ characteristic to be compared with a given experimental characteristic is often deduced directly from the experimental high-frequency $C-V$ data. Let us explore the determination procedure.

(a) The device yielding the high frequency $C-V$ characteristic shown in Fig. 2.17 exhibited a maximum capacitance (C_O) of 82 pf. The gate area of the MOS-C was equal to 4.75×10^{-3} cm^2 and $K_O = 3.9$. Determine x_o from the given data.

(b) Referring to Fig. 2.17, note the minimum high-frequency value of C/C_O observed under far-inversion biasing ($V_G < -4$ V). Per Eq. (C.1), found in Appendix C, one can associate a W_{eff} with the observed C/C_O. Calculate W_{eff}(far-inv) employing $K_S = 11.8$, $K_O = 3.9$, and the x_o determined in part (a). Divide W_{eff}(far-inv) by $L_D = 2.68 \times 10^{-3}$ cm to obtain the experimental value of W_{eff}(far-inv)$/L_D$. (The intrinsic Debye length, L_D, is defined in Appendixes B and C. The value quoted here is appropriate for $T = 300$ K.)

(c) U. Hartmann and P. Schley, *Phys. Stat. Sol. (a)*, **42**, 667 (1977) established that

$$\frac{W_{\text{eff}}(\text{far-inv})}{L_D} = 2e^{-|U_F|/2}\left[17.28 + 99.48 \tanh\left(\frac{|U_F| - 7.936}{47.13}\right)\right]^{1/2}$$

where

$$U_F = \frac{\phi_F}{kT/q}$$

The accuracy of the above relationship is reported to be better than $\pm 0.1\%$ over the range $4 \leq |U_F| \leq 16$. By making a plot of $W_{\text{eff}}(\text{far-inv})/L_D$ versus U_F, through interative techniques using a computer, or by a simple hit-and-miss method, determine the value of U_F needed to match the part (b) experimental result for $W_{\text{eff}}(\text{far-inv})/L_D$. Determine U_F to four significant figures and compute the corresponding N_D assuming $T = 300$ K. [*Note:* An excellent first guess for N_D can be rapidly deduced from Fig. 2.9 by equating $W_{\text{eff}}(\text{far-inv})$ and W_T. The deduced N_D can in turn be used to compute a first guess for U_F.]

3 / MOSFETs–An Introduction

MOSFET-based integrated circuits have become the dominant technology in the semi-conductor industry. There are literally hundreds of MOS-transistor circuits in production today, ranging from rather simple logic gates used in digital-signal processing to custom designs with both logic and memory functions on the same silicon chip. MOS products are found in a mind-boggling number of electronic systems including the now commonplace microcomputer. Initially the MOS-transistor was identified by several competing acronyms; namely, metal–oxide–semiconductor transistor (MOST), insulated gate field effect transistor (IGFET), and metal–oxide–semiconductor field effect transistor (MOSFET). (PIGFET and MISFET were even suggested with a smile at one time or another.) With the passage of time, however, the transistor structure has commonly come to be known as the MOSFET. In this chapter we are concerned with describing the operation of the MOSFET and modeling the device characteristics. We continue to assume the MOS structure to be ideal. Moreover, the development focuses on the basic transistor configuration, the long-channel (or large-dimension) enhancement-mode MOSFET. An examination of small-dimension effects and structural variations is undertaken in Chapter 5. We begin here with a qualitative discussion of MOSFET operation and dc current flow inside the structure, progress through a quantitative analysis of the dc $(I_D–V_D)$ characteristics, and conclude with an examination of the ac response.

3.1 QUALITATIVE THEORY OF OPERATION

Figure 3.1 shows a cross-sectional view of the basic MOS-transistor configuration. Physically, the MOSFET is essentially nothing more than an MOS-capacitor with two pn junctions placed immediately adjacent to the region of the semiconductor controlled by the MOS-gate. The Si substrate can be either p-type (as pictured) or n-type; p^+ junction islands are of course required in n-bulk devices. Also shown in Fig. 3.1 are the standard terminal and dc voltage designations. The drain current (I_D), which flows in response to the applied terminal voltages, is the primary dc observable. Consistent with the naming of the device leads, the current flow is always such that carri-

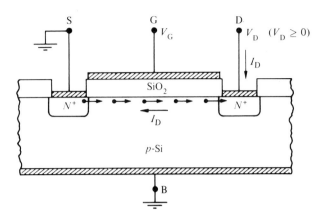

Fig. 3.1   Cross-sectional view of the basic p-bulk (n-channel) MOSFET structure showing the terminal designations, carrier and current flow directions, and standard biasing conditions.

ers (electrons in the present case) enter the structure through the source (S), leave through the drain (D), and are subject to the control or gating action of the gate (G). The voltage applied to the gate relative to ground is V_G, while the drain voltage relative to ground is V_D. Unless stated otherwise, we will assume that the source and back are grounded. Please note that under normal operational conditions the drain bias is always such as to reverse bias the drain pn junction ($V_D \geq 0$ for the Fig. 3.1 device). Finally, reflecting the normal flow of electrons within the structure, the drain current for the p-bulk device is taken to be positive when flowing from the external circuit into the drain terminal.

To determine how the drain current is expected to vary as a function of the applied terminal voltages, let us first conceptually set $V_D = 0$ and examine the situation inside the structure as a function of the imposed gate voltage. When V_G is accumulation or depletion biased ($V_G \leq V_T$, where V_T is the depletion–inversion transition-point voltage), the gated region between the source and drain islands contains either an excess or deficit of holes and very few electrons. Thus, looking along the surface between the n^+ islands under the cited conditions one effectively sees an open circuit. When V_G is inversion biased ($V_G > V_T$), however, an inversion layer containing mobile electrons is formed adjacent to the Si surface. Now looking along the surface between the n^+ islands one sees, as pictured in Fig. 3.2(a), an induced "n-type" region (the inversion layer) or conducting *channel* connecting the source and drain islands. Naturally, the greater the inversion bias, the greater the pile-up of electrons at the Si surface and the greater the conductance of the inversion layer. An inverting gate bias, therefore, creates or induces a source-to-drain channel and determines the maximum conductance of the channel.

Turning next to the action of the drain bias, suppose an inversion bias $V_G > V_T$ is applied to the gate and the drain voltage is increased in small steps starting from $V_D = 0$. At $V_D = 0$ the situation inside the device is as previously pictured in Fig. 3.2(a),

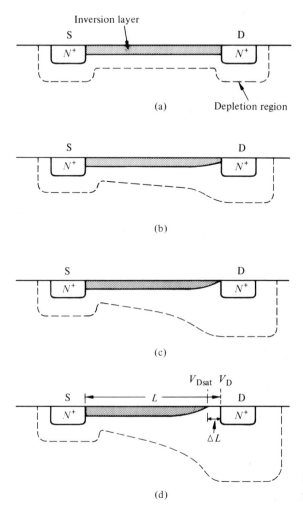

Fig. 3.2   Visualization of various phases of $V_G > V_T$ MOSFET operation. (a) $V_D = 0$; (b) channel (inversion layer) narrowing under moderate V_D biasing; (c) pinch-off; and (d) postpinch-off ($V_D > V_{Dsat}$) operation. (Note that the inversion layer widths, depletion widths, etc. are not drawn to scale.)

thermal equilibrium obviously prevails, and the drain current is identically zero. With V_D stepped to small positive voltages, the surface channel merely acts like a simple resistor and a drain current proportional to V_D begins to flow into the drain terminal. The portion of the I_D–V_D relationship corresponding to small V_D biases is shown as the line from the origin to point A in Fig. 3.3. Any $V_D > 0$ bias, it should be interjected, simultaneously reverse biases the drain pn junction, and the resulting reverse bias

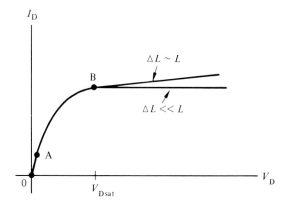

Fig. 3.3   General variation of I_D with V_D for a given $V_G > V_T$.

junction current flowing into the Si substrate does contribute to I_D. In well-made devices, however, the junction leakage current is totally negligible compared to the channel current, provided V_D is less than the junction breakdown voltage.

Once V_D is increased above a few tenths of a volt, the device enters a new phase of operation. Specifically, the voltage drop from the drain to the source associated with the flow of current in the channel starts to negate the inverting effect of the gate. As pictured in Fig. 3.2(b), the depletion region widens in going down the channel from the source to the drain and the number of inversion layer carriers correspondingly decreases. The reduced number of carriers decreases the channel conductance, which in turn is reflected as a decrease in the slope of the observed I_D–V_D characteristic. Continuing to increase the drain voltage causes an ever-increasing depletion of the channel and the systematic slope-over in the I_D–V_D characteristic noted in Fig. 3.3. The greatest decrease in channel carriers occurs of course near the drain, and eventually the inversion layer completely vanishes in the near vicinity of the drain [see Fig. 3.2(c)]. The disappearance of the conducting channel adjacent to the drain in the MOSFET is referred to as "*pinch-off.*" When the channel pinches off inside the device, the point B is reached on the Fig. 3.3 characteristic; that is, the slope of the I_D–V_D characteristic becomes approximately zero.

For drain voltages in excess of the pinch-off voltage, V_{Dsat}, the pinched-off portion of the channel widens from just a point into a depleted channel section ΔL in extent [see Fig. 3.2(d)]. Being a depletion region, the pinched-off ΔL section absorbs most of the voltage drop in excess of V_{Dsat}. Given a long-channel device where $\Delta L \ll L$, the source to pinch-off region of the MOSFET will be essentially identical in shape and will have the same endpoint voltages for all $V_D \geq V_{Dsat}$. When the shape of a conducting region and the potential applied across the region do not change, the current through the region must also remain invariant. Thus, I_D remains approximately constant for drain voltages in excess of V_{Dsat} provided $\Delta L \ll L$. If ΔL is comparable to L, the same voltage drop (V_{Dsat}) will appear across a shorter channel ($L - \Delta L$) and, as noted in Fig. 3.3, the post-pinch-off I_D in such devices will increase somewhat with increasing $V_D > V_{Dsat}$.

Thus far we have examined the response of the MOSFET to the separate manipulation of the gate and drain biases. To establish a complete set of I_D–V_D characteristics it is necessary to combine the results derived from the separate considerations. Clearly, for $V_G \leq V_T$, the gate bias does not create a surface channel and $I_D \simeq 0$ for all drain biases below the junction breakdown voltage. For all $V_G > V_T$ a characteristic of the form shown in Fig. 3.3 will be observed. Since the conductance of the channel increases with increasing V_G, it follows that the initial slope of the I_D–V_D characteristic will likewise increase with increasing V_G. Moreover, the greater the number of inversion layer carriers present when $V_D = 0$, the larger the drain voltage required to achieve pinch-off. Thus V_{Dsat} must increase with increasing V_G. From the foregoing arguments one concludes that the variation of I_D with V_D and V_G must be of the form displayed in Fig. 3.4.

The I_D–V_D characteristics just established confirm the "transistor" nature of the MOSFET structure. Transistor action results, of course, when I_D flowing in an output circuit is modulated by an input voltage applied to the gate. Relative to terminology, we should note that the portion of the characteristics where $V_D > V_{Dsat}$ for a given V_G is referred to as the *saturation* region of operation; the portion of the characteristics where $V_D < V_{Dsat}$ is called the *linear* (or sometimes *triode*) region of operation. Also, the MOSFET is known as an *n-channel* device when the channel carriers are electrons; when the channel carriers are holes the MOSFET is designated a *p-channel* device.

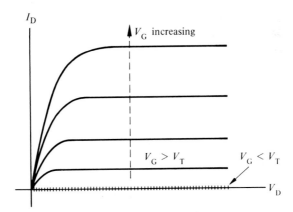

Fig. 3.4 General form of the I_D–V_D characteristics expected from a long channel ($\Delta L \ll L$) MOSFET.

3.2   QUANTITATIVE I_D–V_D RELATIONSHIPS

During the course of MOSFET development there has evolved a hierarchy of long-channel I_D–V_D formulations that provide progressively increased accuracy at the expense of increased complexity. We will examine two of the formulations: namely, the "square-law" theory and the "bulk-charge" theory. The former provides very simple relationships; the latter is a much more accurate representation of reality. Interestingly, all but the final derivational steps in the two theories are identical. Comments relative to more exacting long-channel theories can be found at the end of the section.

3.2.1   Preliminary Considerations

Threshold Voltage

From the qualitative description of MOSFET operation it should be obvious that the parameter V_T plays a prominent role in determining the precise nature of the device characteristics. In MOSFET analyses V_T is commonly called the *threshold* or *turn-on* voltage. The transistor starts to carry current (turns on) at the onset of inversion. A computational expression for the important V_T parameter is readily established using the results of Subsection 2.3.2 and the fact that $V_G = V_T$ when $\phi_S = 2\phi_F$. Specifically, given an ideal *n*-channel (or *p*-bulk) device, simple substitution into Eq. (2.28) yields

$$V_T = 2\phi_F + \frac{K_S x_o}{K_O} \sqrt{\frac{4qN_A}{K_S \varepsilon_0} \phi_F} \qquad \begin{array}{l} \text{... ideal } n\text{-channel} \\ (p\text{-bulk) devices} \end{array} \qquad (3.1a)$$

Analogously,

$$V_T = 2\phi_F - \frac{K_S x_o}{K_O} \sqrt{\frac{4qN_D}{K_S \varepsilon_0}(-\phi_F)} \qquad \begin{array}{l} \text{... ideal } p\text{-channel} \\ (n\text{-bulk) devices} \end{array} \qquad (3.1b)$$

Effective Mobility

In deriving quantitative expressions for the MOSFET dc characteristics one encounters a new parameter known as the "effective mobility." The carrier mobilities, μ_n and μ_p, were first described in Section 3.1 of Volume I and were noted to be a measure of the ease of carrier motion within a semiconductor crystal. In the semiconductor bulk, that is, at a point far removed from the semiconductor surface, the carrier mobilities are typically determined by the amount of lattice scattering and ionized impurity scattering taking place inside the material. For a given temperature and semiconductor doping, these bulk mobilities (μ_n and μ_p) are well-defined and well-documented material constants. Carrier motion in a MOSFET, however, takes place in a surface-inversion layer where the gate-induced electric field acts so as to accelerate the carriers toward the surface. The inversion layer carriers therefore experience motion impeding collisions with the Si surface (see Fig. 3.5) in addition to lattice and ionized impurity scattering. The additional surface scattering mechanism lowers the mobility of the carriers, with the carriers constrained nearest the Si surface experiencing the greatest reduction in mobility. The resulting average mobility of the inversion layer carriers is called the *effective mobility* and is given the symbol $\overline{\mu}_n$ or $\overline{\mu}_p$.

Seeking to establish a formal mathematical expression for the effective mobility, let us consider an *n*-channel device with the structure and dimensions specified in Fig. 3.6. Let x be the depth into the semiconductor measured from the oxide–semiconductor interface, y the distance along the channel measured from the source, $x_c(y)$ the channel depth, $n(x, y)$ the electron concentration at a point (x, y) in the channel, and $\mu_n(x, y)$ the mobility of carriers at the (x, y) point in the channel. Invoking the standard averaging procedure, the effective mobility of carriers an arbitrary distance y from the source is simply given by

$$\overline{\mu}_n = \frac{\displaystyle\int_0^{x_c(y)} \mu_n(x, y)n(x, y)\, dx}{\displaystyle\int_0^{x_c(y)} n(x, y)\, dx} \tag{3.2}$$

However, the total electronic charge/cm^2 in the channel at a given y is

$$Q_N(y) = -q \int_0^{x_c(y)} n(x, y)\, dx \tag{3.3}$$

Consequently, one can alternatively write

$$\overline{\mu}_n = -\frac{q}{Q_N(y)} \int_0^{x_c(y)} \mu_n(x, y)n(x, y)\, dx \tag{3.4}$$

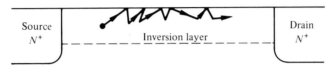

Source
N^+

Inversion layer

Drain
N^+

Fig. 3.5 Visualization of surface scattering at the Si–SiO$_2$ interface.

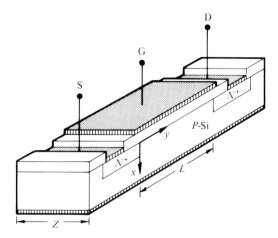

Fig. 3.6   Device structure, dimensions, and coordinate orientations assumed in the quantitative analysis.

If the drain voltage is small, the channel depth and carrier charge will be more or less uniform from source to drain and the effective mobility will be essentially the same for all y values. When the drain voltage becomes large, on the other hand, x_c and Q_N vary with position, and it is reasonable to expect that $\bar{\mu}_n$ likewise varies somewhat in going down the channel from the source to the drain. Fortunately, the cited y-dependence can be neglected without introducing a significant error. Thus, *herein we will subsequently consider $\bar{\mu}_n$ to be a device parameter that is independent of y and the applied drain voltage V_D.*

Relative to the dependence of $\bar{\mu}_n$ on the applied *gate* voltage, increased inversion biasing places more carriers closer to the oxide–semiconductor interface and increases the electric field acting on the carriers. This combination of effects enhances surface scattering and thereby lowers the average carrier mobility; $\bar{\mu}_n$ therefore decreases with increased inversion biasing — a dependence that cannot be ignored. The exact $\bar{\mu}_n$ versus V_G dependence varies from device to device but generally follows the form displayed in Fig. 3.7. Also note from Fig. 3.7 that the surface scattering phenomenon can be rather significant, giving rise to effective mobilities considerably below the bulk μ_n.

3.2.2   Square-Law Theory

The MOSFET under analysis is taken to be a long-channel device with the structure and dimensions as specified in Fig. 3.6. The figure also indicates the assumed coordinate orientations.

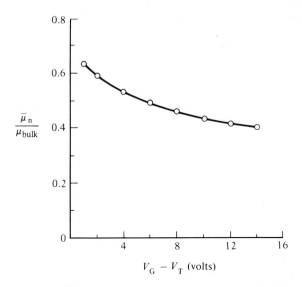

Fig. 3.7 Sample variation of $\bar{\mu}_n$ with the applied gate voltage ($V_D \cong 0$). [Data from S. C. Sun and J. D. Plummer, *IEEE Trans. on E.D.*, **ED-27**, 1497 (August 1980).]

For gate voltages above turn-on ($V_G \geq V_T$), and drain voltages below pinch-off ($0 \leq V_D \leq V_{Dsat}$), the derivation of the square-law I_D–V_D relationship proceeds as follows: In general one can write

$$\mathbf{J_N} = q\mu_n n\boldsymbol{\mathscr{E}} + qD_N\nabla n \qquad (3.5)$$

Within the conducting channel the current flow is almost exclusively in the y-direction. Moreover, the diffusion component of the current is often found to be negligible when dealing with the more numerous carrier at a given point inside a semiconductor. Thus, based on precedent established in similar problems, it is reasonable to neglect the diffusion component of the current ($qD_N\nabla n$) in Eq. (3.5). Implementing the suggested simplifications yields

$$J_N \cong J_{Ny} \cong q\mu_n n\mathscr{E}_y = -q\mu_n n\frac{d\phi}{dy} \qquad \text{(in the conducting channel)} \qquad (3.6)$$

All of the quantities in Eq. (3.6)—μ_n, n, and J_{Ny}—are, of course, x- and y-position dependent. J_{Ny}, like n, is expected to be quite large at $x = 0^+$ and to drop off rapidly as one moves into the semiconductor bulk.

Since current flow is restricted to the surface channel, the current passing through any cross-sectional plane within the channel must be equal to I_D. That is,*

$$I_D = -\iint J_{Ny}\, dx\, dz = -Z \int_0^{x_c(y)} J_{Ny}\, dx \tag{3.7a}$$

$$= \left(-Z\frac{d\phi}{dy}\right)\left(-q\int_0^{x_c(y)} \mu_n(x, y)n(x, y)\, dx\right) \tag{3.7b}$$

Upon examining Eq. (3.7b), note that the second bracket on the right-hand side of the equation is just $\bar{\mu}_n Q_N$ [see Eq. (3.4)]. Eq. (3.7b) thus simplies to

$$I_D = -Z\bar{\mu}_n Q_N \frac{d\phi}{dy} \tag{3.8}$$

Next, realizing that I_D is independent of y, we can recast Eq. (3.8) into a more useful form by integrating I_D over the length of the channel. Specifically,

$$\int_0^L I_D\, dy = I_D L = -Z \int_0^{V_D} \bar{\mu}_n Q_N\, d\phi \tag{3.9}$$

or, with $\bar{\mu}_n$ being position independent,

$$I_D = -\frac{Z\bar{\mu}_n}{L} \int_0^{V_D} Q_N\, d\phi \tag{3.10}$$

An analytical expression relating Q_N to the channel potential ϕ at an arbitrary point y is obviously required to complete the derivation. Working to establish the required expression, we recall that the equilibrium inversion-layer charge in an MOS-C almost precisely balances the charge added to the MOS-C gate when V_G exceeds V_T. In other words

$$\Delta Q_{gate}\left(\frac{charge}{cm^2}\right) = -\Delta Q_{semi}\left(\frac{charge}{cm^2}\right) \cong -Q_N \quad \dots V_G \geq V_T \tag{3.11}$$

Because the charges are added immediately adjacent to the edges of the oxide, we can also assert

$$\Delta Q_{gate}\left(\frac{charge}{cm^2}\right) \cong C_o \Delta V_G = C_o(V_G - V_T) \quad \dots V_G \geq V_T \tag{3.12}$$

---

*(a) The minus sign appears in the general formula for I_D because I_D is defined to be positive in the $-y$-direction. (b) Generally speaking, ϕ and $d\phi/dy$ are functions of x. The very narrow extent of the inversion layer dictates, however, that ϕ is a weak function of x ($\phi \approx \phi_S$) in the channel region. Therefore, in writing down the final form of Eq. (3.7), $d\phi/dy$ was taken to be constant over the x-width of the channel.

and therefore

$$Q_N \cong -C_o(V_G - V_T) \qquad \ldots V_G \geq V_T \tag{3.13}$$

where

$$C_o \equiv \frac{C_O}{A_G} = \frac{K_O \varepsilon_0}{x_o}$$

is the oxide capacitance per unit area of the gate.

Whereas the entire back side in an MOS-C is grounded, the bottom-side "plate" potential in a MOSFET varies from zero at the source to V_D at the drain. As envisioned in Fig. 3.8, the MOSFET can be likened to a resistive-plate capacitor where the plate-to-plate potential difference is V_G at the source, $V_G - V_D$ at the drain, and $V_G - \phi$ at an arbitrary point y. Clearly, the potential drop $V_G - \phi$ at an arbitrary point y in the MOSFET functionally replaces the uniform V_G potential drop in an MOS-C. Utilizing Eq. (3.13) we therefore conclude

$$Q_N(y) \cong -C_o(V_G - V_T - \phi) \tag{3.14}$$

An explicit I_D–V_D relationship can now be established by simply substituting the Eq. (3.14) expression for Q_N into Eq. (3.10) and integrating. The result is

$$I_D = \frac{Z\bar{\mu}_n C_o}{L}\left[(V_G - V_T)V_D - \frac{V_D^2}{2}\right] \quad \begin{pmatrix} 0 \leq V_D \leq V_{Dsat} \\ V_G \geq V_T \end{pmatrix} \tag{3.15}$$

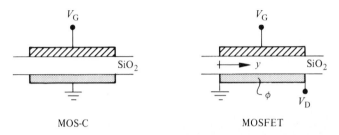

MOS-C   MOSFET

Fig. 3.8 Capacitor-like model for determining the charge in the MOSFET channel.

It should be reemphasized that the foregoing development and Eq. (3.15), in particular, apply only below pinch-off. In fact, the computed I_D versus V_D for a given V_G actually begins to decrease if V_D values in excess of V_{Dsat} are inadvertently substituted into Eq. (3.15). As pointed out in the qualitative discussion, however, I_D is approximately constant if V_D exceeds V_{Dsat}. To first order, then, the post pinch-off portion of the characteristics can be modeled by simply setting

$$I_{D|V_D>V_{Dsat}} = I_{D|V_D=V_{Dsat}} \equiv I_{Dsat} \tag{3.16}$$

or

$$I_{Dsat} = \frac{Z\overline{\mu}_n C_o}{L}\left[(V_G - V_T)V_{Dsat} - \frac{V_{Dsat}^2}{2}\right] \tag{3.17}$$

The I_{Dsat} relationship can be simplified somewhat by noting that pinch-off at the drain end of the channel implies $Q_N(L) \to 0$ when $\phi(L) = V_D \to V_{Dsat}$. Thus from Eq. (3.14)

$$Q_N(L) = -C_o(V_G - V_T - V_{Dsat}) = 0 \tag{3.18}$$

or

$$\boxed{V_{Dsat} = V_G - V_T} \tag{3.19}$$

and

$$\boxed{I_{Dsat} = \frac{Z\overline{\mu}_n C_o}{2L}(V_G - V_T)^2} \tag{3.20}$$

Neglecting $\overline{\mu}_n$'s dependence on V_G, Eq. (3.20) predicts a saturation drain current that varies as the square of the gate voltage above turn-on, the so-called "square-law" dependence.

SEE EXERCISE 3.1 — APPENDIX A

3.2.3 Bulk-Charge Theory

Although appearing very reasonable and sound on first inspection, close scrutiny reveals that the square-law theory contains a major flaw. The capacitorlike model used in the square-law analysis assumed changes in gate charge going down the MOSFET channel were balanced solely by changes in Q_N. This is equivalent to implicitly assuming the depletion width at all channel points from the source to the drain remains fixed at W_T even under $V_D \neq 0$ biasing. In reality, as pictured in Figs. 3.2(b) to (d), the depletion width widens in progressing from the source to the drain when $V_D \neq 0$. This point-to-point variation in the depletion layer or "bulk" charge must be included in any charge balance relationship.

With changes in the depletion width, $W(y)$, taken into account, one more accurately deduces

$$Q_N(y) = -C_o(V_G - V_T - \phi) + qN_A[W(y) - W_T] \tag{3.21}$$

where, making use of the delta-depletion results in Chapter 2,

$$W(y) = \left[\frac{2K_S\varepsilon_0}{qN_A}(2\phi_F + \phi)\right]^{1/2} \tag{3.22}$$

$$W_T = \left[\frac{2K_S\varepsilon_0}{qN_A}(2\phi_F)\right]^{1/2} \tag{3.23}$$

Thus, combining Eqs. (3.21) to (3.23) and introducing

$$V_W \equiv \frac{qN_A W_T}{C_o} \tag{3.24}$$

one obtains the bulk-charge theory analogue of Eq. (3.14), namely,

$$Q_N(y) = -C_o\left[V_G - V_T - \phi - V_W\left(\sqrt{1 + \frac{\phi}{2\phi_F}} - 1\right)\right] \tag{3.25}$$

The predicted I_D-V_D relationship based on the bulk-charge formulation is next readily obtained by substituting Eq. (3.25) into Eq. (3.10) and integrating. The end result is

$$I_D = \frac{Z\overline{\mu}_n C_o}{L}\left\{(V_G - V_T)V_D - \frac{V_D^2}{2} - \frac{4}{3}V_W\phi_F\left[\left(1 + \frac{V_D}{2\phi_F}\right)^{3/2} - \left(1 + \frac{3V_D}{4\phi_F}\right)\right]\right\}$$

$$\text{for } 0 \leq V_D \leq V_{Dsat}$$
$$\text{and } V_G \geq V_T \tag{3.26}$$

As in the square-law analysis the post pinch-off portion of the characteristics are approximately modeled by setting I_D evaluated at $V_D > V_{Dsat}$ equal to I_D at $V_D = V_{Dsat}$. Likewise, an expression for V_{Dsat} can be obtained by noting $Q_N(y)|_{y=L} \to 0$ in Eq. (3.25) when $\phi(L) = V_D \to V_{Dsat}$. One finds

$$V_{Dsat} = V_G - V_T - V_W \left\{ \left[\frac{V_G - V_T}{2\phi_F} + \left(1 + \frac{V_W}{4\phi_F} \right)^2 \right]^{1/2} - \left(1 + \frac{V_W}{4\phi_F} \right) \right\} \qquad (3.27)$$

Having concluded the mathematical development, let us examine the results and make appropriate comments. First of all, it should be recognized that the primary asset of the square-law theory is its simplicity. General trends, basic interrelationships, and the like can be established using the square-law formulation without an excessive amount of mathematical entanglement. The bulk-charge theory, on the other hand, is in good agreement with the experimental characteristics derived from long-channel MOSFETs. It should also be noted that, although Eqs. (3.26) and (3.27) are decidedly more complex than their square-law analogues, the added terms, the terms not appearing in Eqs. (3.15) and (3.19), respectively, are always negative and act primarily to reduce I_D and V_{Dsat} for a given set of operational conditions. Figure 3.9, which compares the two theories, confirms the foregoing observation and also illustrates another well-known

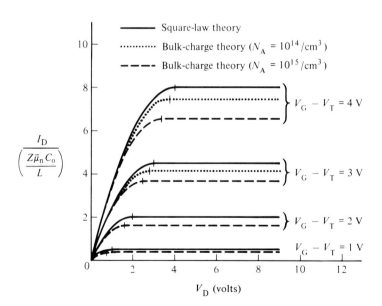

Fig. 3.9 Comparison of the I_D–V_D characteristics derived from the square-law and bulk-charge theories. The bulk-charge curves were computed assuming $x_o = 0.1 \ \mu m$ and $T = 300$ K.

property—the accuracy of the square-law theory improves as the substrate doping is decreased. In fact, the bulk-charge theory mathematically reduces to the square-law theory as N_A (or N_D) $\rightarrow 0$ and $x_o \rightarrow 0$.

Relative to more-exacting long-channel theories, it should be noted that both the square-law and bulk-charge theories suffer from two severe inherent limitations. For one, the drain current is assumed to be identically zero for gate biases below turn-on. Although small, the "subthreshold" drain current is not identically zero and its precise value is sometimes of interest. Second, the square-law and bulk-charge relationships do not self-saturate—one must artificially construct the postpinch-off portion of the characteristics. The cited failings are removed in the *charge-sheet* and *exact-charge* formulations; either theory can be used to compute the subthreshold current and both theories are self-saturating. The I_D-V_D computational relationships resulting from the charge-sheet and exact-charge models are reproduced in Appendix D. The exact-charge results are of course established by working with the exact-charge distribution inside the MOSFET. Although not overly complex, the exact-charge result does involve integrals. The charge-sheet model may be viewed as a simplified version of the exact formulation. Of the theories discussed, the charge-sheet formulation provides the best trade-off between accuracy and complexity. Sample current-voltage and subthreshold characteristics constructed using the cited more-exacting theories are shown in Figs. 3.10 and 3.11, respectively.

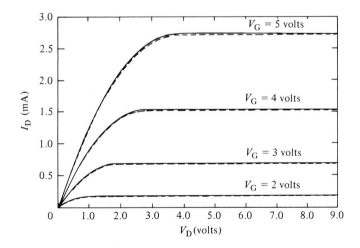

Fig. 3.10 Theoretical current-voltage characteristics of an *n*-channel MOSFET with $x_o = 0.05\ \mu\text{m}$, $N_A = 10^{15}/\text{cm}^3$, $\bar{\mu}_n = 550\ \text{cm}^2/\text{V-sec}$, $L = 7\ \mu\text{m}$, $Z = 70\ \mu\text{m}$ and $T = 23°\text{C}$. The solid-line curves were derived from the exact-charge result while the dashed-line curves were computed using the charge-sheet theory. (Reprinted with permission from *Solid-State Electronics,* R. F. Pierret and J. A. Shields, **26,** 143, © 1983, Pergamon Press plc.)

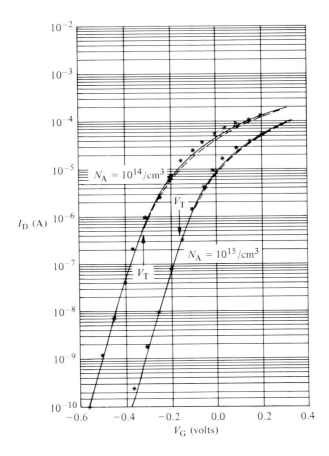

Fig. 3.11 Subthreshold transfer characteristics of n-channel MOSFETs having the same parameters as the Fig. 3.10 device except $N_A = 10^{14}/cm^3$ or $N_A = 10^{15}/cm^3$ and $x_o = 0.013\ \mu m$. The solid- and dashed-line curves were computed respectively from the exact-charge and charge-sheet results with $V_D = 1$ V. The (*) are experimental data. (Reprinted with permission from *Solid-State Electronics*, **26**, R. F. Pierret and J. A. Shields, © 1983, Pergamon Press plc.)

3.3   ac RESPONSE

3.3.1   Small Signal Equivalent Circuits

The ac response of the MOSFET, routinely expressed in terms of a small-signal equivalent circuit, is most conveniently established by considering the two-port network shown in Fig. 3.12(a). Initially we restrict our considerations to low operational fre-

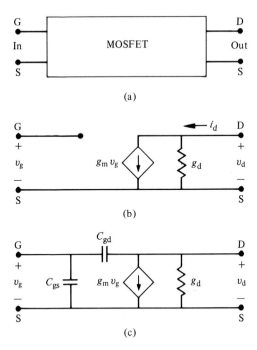

Fig. 3.12   (a) The MOSFET viewed as a two-port network. (b) Low-frequency and (c) high-frequency small-signal equivalent circuits characterizing the ac response of the MOSFET.

quencies where capacitive effects may be neglected. It should be noted that the following development and results are very similar to those of the J-FET presentation in Section 1.4.

We begin by examining the device input. Looking into the input port between the gate and grounded source/substrate one sees a capacitor. A capacitor, however, behaves (to first order) like an open circuit at low frequencies. It is standard practice, therefore, to model the low-frequency input to the MOSFET by an open circuit.

At the output port the dc drain current has been established to be a function of V_D and V_G; that is, $I_D = I_D(V_D, V_G)$. When ac drain and gate potentials, v_d and v_g, are respectively added to the dc drain and gate terminal voltages, V_D and V_G, the drain current through the structure is of course modified to $I_D(V_D, V_G) + i_d$, where i_d is the ac component of the drain current. Provided the device can follow the ac changes in potential, which is certainly the case at low operational frequencies, one can state

$$i_d + I_D(V_D, V_G) = I_D(V_D + v_d, V_G + v_g) \tag{3.28a}$$

and

$$i_d = I_D(V_D + v_d, V_G + v_g) - I_D(V_D, V_G) \tag{3.28b}$$

Expanding the first term on the right-hand side of Eq. (3.28b) in a Taylor series and keeping only through first-order terms in the expansion (higher-order terms are negligible), one obtains

$$i_d = \left.\frac{\partial I_D}{\partial V_D}\right|_{V_G} v_d + \left.\frac{\partial I_D}{\partial V_G}\right|_{V_D} v_g \tag{3.29a}$$

or

$$i_d = g_d v_d + g_m v_g \tag{3.29b}$$

where

$$g_d \equiv \left.\frac{\partial I_D}{\partial V_D}\right|_{V_G = \text{constant}} \qquad \text{the drain or channel conductance} \tag{3.30a}$$

$$g_m \equiv \left.\frac{\partial I_D}{\partial V_G}\right|_{V_D = \text{constant}} \qquad \text{transconductance or mutual conductance} \tag{3.30b}$$

Equation (3.29b) may be viewed as the ac-current node equation for the drain terminal and, by inspection, leads to the output portion of the circuit displayed in Fig. 3.12(b). Since, as concluded earlier, the gate-to-source or input portion of the device is simply an open circuit, Fig. 3.12(b) then is the desired small-signal equivalent circuit characterizing the low-frequency ac response of the MOSFET.

For field-effect transistors the g_m parameter plays a role analogous to the α's and β's in the modeling of bipolar junction transistors. As its name indicates, g_d may be viewed as either the device output admittance or the ac conductance of the channel between the source and drain. Explicit g_d and g_m relationships obtained by direct differentiation of Eqs. (3.15), (3.20), and (3.26) using the Eq. (3.30) definitions are catalogued in Table 3.1.

At the higher operational frequencies often encountered in practical applications, the Fig. 3.12(b) circuit must be modified to take into account capacitive coupling between the device terminals. The required modification is shown in Fig. 3.12(c). A capacitor between the drain and source terminals at the output has been omitted in Fig. 3.12(c) because the drain-to-source capacitance is typically negligible. C_{gd}, which provides undesirable feedback between the input and output, is associated in large part with the so-called overlap capacitance — the capacitance resulting from the portion of

Table 3.1   MOSFET Small Signal Parameters.*

	Below pinch-off ($V_D \le V_{Dsat}$)	Post pinch-off ($V_D > V_{Dsat}$)
Square law	$g_d = \dfrac{Z\bar{\mu}_n C_o}{L}(V_G - V_T - V_D)$	$g_d = 0$
Bulk charge	$g_d = \dfrac{Z\bar{\mu}_n C_o}{L}[V_G - V_T - V_D$ $- V_W(\sqrt{1 + V_D/2\phi_F} - 1)]$	$g_d = 0$
Square law	$g_m = \dfrac{Z\bar{\mu}_n C_o}{L}V_D$	$g_m = \dfrac{Z\bar{\mu}_n C_o}{L}(V_G - V_T)$
Bulk charge	$g_m = \dfrac{Z\bar{\mu}_n C_o}{L}V_D$	$g_m = \dfrac{Z\bar{\mu}_n C_o}{L}V_{Dsat}$ with V_{Dsat} per Eq. (3.27)

*Entries in the table were obtained by direct differentiation of Eqs. (3.15), (3.20), and (3.26). The variation of $\bar{\mu}_n$ with V_G was neglected in establishing the g_m expressions.

the gate that overlaps the drain island. The overlap capacitance is minimized by forming a thicker oxide in the overlap region or preferably through the use of self-aligned gate procedures. In the self-aligned gate fabrication process a MOSFET gate material that can withstand high-temperature processing, usually poly-Si, is deposited first. After the gate is defined, the source and drain islands are subsequently formed abutting the gate by diffusion or ion implantation. The remaining capacitor shown in Fig. 3.12(c), C_{gs}, is of course associated primarily with the capacitance of the MOS gate.

3.3.2   Cutoff Frequency

Given the small-signal equivalent circuit of Fig. 3.12(c) it is possible to estimate the maximum operating frequency or cutoff frequency of an MOS transistor. Let f_{max} be defined as the frequency where the MOSFET is no longer amplifying the input signal under optimum conditions; that is, the frequency where the absolute value of the output current to input current ratio is unity when the output of the transistor is short-circuited. By inspection, the input current with the output short-circuited is

$$i_{in} = j\omega(C_{gs} + C_{gd})v_g \simeq j(2\pi f)C_o v_g \qquad (j = \sqrt{-1}) \qquad (3.31)$$

where C_{gd} is taken to be small and $C_{gs} \simeq C_o$. Likewise, the output current is

$$i_{out} \simeq g_m v_g \qquad (3.32)$$

Thus, setting $|i_{out}/i_{in}| = 1$ and solving for $f = f_{max}$, one obtains

$$f_{max} = \frac{g_m}{2\pi C_O} = \frac{\bar{\mu}_n V_D}{2\pi L^2} \qquad \text{if } V_D \leq V_{Dsat} \qquad (3.33)$$

The latter form of Eq. (3.33) was established using the below pinch-off g_m entry in Table 3.1. The important point to note is that the channel length L is the key parameter in determining f_{max}; increased MOSFET operating frequencies are achieved by decreasing the channel length.

3.3.3 Small-Signal Characteristics

Representative sketches of selected small-signal characteristics that have received special attention in the device literature are shown in Fig. 3.13. g_d versus V_G with $V_D = 0$ has been used to obtain a reasonably accurate estimate of V_T. This is accomplished by extrapolating the linear portion of the g_d–V_G characteristics into the V_G axis and equating the voltage intercept to V_T. The basis for this procedure can be understood by referring to the below pinch-off g_d entries in Table 3.1. With $V_D = 0$ the drain conductance in both the square-law and bulk-charge theories reduces to

$$g_d = \frac{Z \bar{\mu}_n C_o}{L}(V_G - V_T) \qquad (V_D = 0) \qquad (3.34)$$

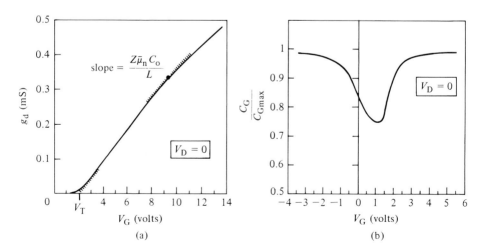

Fig. 3.13   MOSFET small signal characteristics. (a) g_d versus V_G with $V_D = 0$; (b) C_G versus V_G with $V_D = 0$.

To first order, then, g_d is predicted to be a linear function of V_G, going to zero when $V_G = V_T$. The experimental characteristic does not completely vanish at $V_G = V_T$ because there is a small minority carrier concentration in the surface channel at the depletion–inversion transition point. This residual concentration is neglected in both the square-law and bulk-charge theories. The g_d versus V_G characteristic with $V_D = 0$ has also been used to deduce the effective mobility. Since g_d is directly proportional to $\overline{\mu}_n$ according to Eq. (3.34), $\overline{\mu}_n$ versus V_G can be obtained readily from the slope of the g_d–V_G characteristic. This mobility measurement method is accurate provided the device contains a low density of interfacial traps (see Subsection 4.2.4). A moderate to large density of interfacial traps would spread out the g_d–V_G characteristic and yield a fallaciously low value for $\overline{\mu}_n$.

The second characteristic in Fig. 3.13 typifies the gate capacitance versus V_G dependence derived from the MOSFET when the drain is grounded. The MOSFET C_G–V_G ($V_D = 0$) characteristic has been used for diagnostic purposes in much the same manner as the MOS-C C–V_G characteristic. The MOSFET characteristic can, in fact, be modeled to first order by the low-frequency MOS-C C–V_G theory. Unlike the MOS-C, however, a low-frequency type characteristic is observed even when the MOSFET is probed at frequencies exceeding 1 MHz. A low-frequency characteristic is obtained because the source and drain islands supply the minority carriers required for the structure to follow the ac fluctuations in the gate potential when the device is inversion biased. Minority carriers merely use the surface channel to flow laterally into and out of the MOS gate area in response to the applied ac signal.

> **SEE EXERCISE 3.2 — APPENDIX A**

3.4 SUMMARY

This chapter was intended to provide an introduction to MOSFET terminology, operation, and analysis. The MOS structure was assumed to be ideal and considerations were limited to the basic transistor configuration. We began with a qualitative discussion of MOSFET operation and dc current flow inside the structure. When biased into inversion, the induced surface inversion layer forms a conducting channel between the source and drain contacts. The greater the applied gate voltage in excess of turn-on, the larger the conductance of the internal channel at a given drain voltage. A nonzero drain voltage in turn initiates current flow between the source and drain. The current flow is proportional to V_D at low drain voltages, slopes-over due to channel narrowing as V_D is increased, and eventually saturates once the terminal channel vanishes or pinches-off near the drain.

The quantitative analysis of the MOSFET dc characteristics, considered next, was subject to two notable complications. First of all, carriers in a surface channel experience motion-impeding collisions with the Si surface, which lower the mobility of the carriers and necessitate the introduction of an effective carrier mobility. Second, the carrier concentration and therefore the current density in the surface channel are strong functions of position, dropping off rapidly as one proceeds into the semiconductor bulk. Nonetheless, the first-order results for the MOSFET current–voltage relationship are surprisingly simple. The results of the first-order theory, referred to herein as the square-law theory, are contained in Eqs. (3.15), (3.19), and (3.20). The bulk-charge theory, a second formulation culminating in Eqs. (3.26) and (3.27), provides a more accurate representation of reality at the expense of complexity. Even more-exacting formulations were discussed briefly, with the reader referred to Appendix D for computational details.

The last section of the chapter was devoted to the ac response of the MOSFET. The equivalent circuits of Figs. 3.12(b) and (c), respectively, specify the small-signal response at low and high frequencies. With the aid of the Fig. 3.12(c) circuit it was established that a short-channel length is necessary to achieve high-frequency, high-speed operation. It was also pointed out that useful information can often be extracted from the small-signal parameters (g_d, g_m, C_G) monitored as a function of the dc terminal voltages.

Although designed to be self-contained, the MOSFET development in the present chapter does closely parallel the J-FET presentation in Chapter 1. The reader may find it a useful exercise to note similarities and differences in the operation and analysis of the two devices. In Chapter 4 we examine the impact of nonidealities on MOSFET operation. As noted previously, an examination of small-dimension effects and structural variations is undertaken in Chapter 5.

PROBLEMS

3.1 Answer the following questions as concisely as possible.

(a) Why are the current carrying contacts in the MOSFET referred to as the "source" and "drain"?

(b) Precisely what is the "channel" in MOSFET terminology?

(c) What is the relationship between the depletion–inversion transition point voltage introduced in the MOS-C discussion and the threshold (turn-on) voltage introduced in the MOSFET discussion?

(d) Why is the mobility in the surface channel of a MOSFET different from the carrier mobility in the semiconductor bulk?

(e) What is the likely origin of the "square-law" and "bulk-charge" names used in identifying I_D–V_D theories?

(f) What is the mathematical definition of the drain conductance? the transconductance?

(g) Why is the observed MOSFET C_G–V_G ($V_D = 0$) curve typically a low-frequency characteristic even at a measurement frequency of 1 MHz?

3.2 Suppose Sections 3.1 and 3.2 are to be rewritten using a p-channel MOSFET for illustrative purposes instead of the presently assumed n-channel device.

(a) Indicate how the figures in Section 3.1 must be modified if the MOSFET used for illustrative purposes is a p-channel device.

(b) Indicate the required revisions to the equations in Subsection 3.2.2 if the square-law analysis is performed on a p-channel device.

3.3 Given an ideal p-channel MOSFET maintained at room temperature:

(a) Assuming $V_D = 0$, sketch the MOS energy band diagram for the gate region of the given transistor at threshold.

(b) Assuming $V_D = 0$, sketch the MOS block charge diagram for the gate region of the given transistor at threshold.

(c) Sketch the inversion layer and depletion region inside the MOSFET at pinch-off. Show and label all parts of the transistor.

3.4 An n-channel MOSFET maintained at $T = 300$ K is characterized by the following parameters: $Z = 50~\mu m$, $L = 5~\mu m$, $x_o = 0.05~\mu m$, $N_A = 10^{15}/cm^3$, and $\bar{\mu}_n = 800~cm^2/V\text{-sec}$ (assumed independent of V_G). Using $K_O = 3.9$, $K_S = 11.8$, $kT/q = 0.0259$ V, and $n_i = 1.18 \times 10^{10}/cm^3$, determine:

(a) V_T;

(b) I_{Dsat} (square-law theory) if $V_G = 2$ V;

(c) I_{Dsat} (bulk-charge theory) if $V_G = 2$ V;

(d) g_d if $V_G = 2$ V and $V_D = 0$;

(e) g_m (square-law theory) if $V_G = 2$ V and $V_D = 2$ V;

(f) g_m (bulk-charge theory) if $V_G = 2$ V and $V_D = 2$ V;

(g) f_{max} if $V_G = 2$ V and V_D and $V_D = 1$ V.

3.5 If Eq. (3.15) is used to compute I_D as a function of V_D for a given V_G, and if V_D is allowed to increase above V_{Dsat}, one finds I_D to be a peaked function of V_D maximizing at V_{Dsat}. The foregoing suggests a second way to establish the Eq. (3.19) relationship for V_{Dsat}. Specifically, show that the standard mathematical procedure for determining extrema points of a function can be used to derive Eq. (3.19) directly from Eq. (3.15).

3.6 Suppose a battery $V_B \geq 0$ is connected between the gate and drain of an ideal n-channel MOSFET as pictured in Fig. P3.6. Using the square-law results,

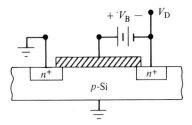

Fig. P3.6

(a) Sketch I_D versus V_D ($V_D \geq 0$) if $V_B = V_T/2$;

(b) Sketch I_D versus V_D ($V_D \geq 0$) if $V_B = 2V_T$.

3.7 In this problem we wish to explore the temperature dependence of the MOSFET dc characteristics. Assume an n-channel MOSFET with $x_o = 0.1\ \mu m$ and $N_A = 10^{16}/cm^3$. Also assume $\bar{\mu}_n$ has the same temperature dependence as μ_n. Make free use of plots in Volume I of the Modular Series.

(a) Compute V_T at $T = 300$ K and at 50° C intervals between $-50°$ C and 200° C.

(b) Using the square-law theory, compute $I_{Dsat}(T)/I_{Dsat}(300$ K) at 50° C intervals between $-50°$ C and 200° C. First set $V_G = 3$ V; repeat the computation taking $V_G = 10$ V.

(c) Discuss your results.

3.8 Derive Eq. (3.27).

3.9 A linear device geometry and a rectangular gate of length L by width Z were explicitly assumed in the text derivation of I_D–V_D relationships. However, MOSFETs have been built with circular geometry as pictured (top view) in Fig. P3.9.

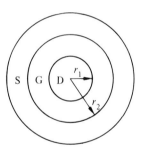

Fig. P3.9

(a) If r_1 and r_2 are the inside and outside diameters of the gated area, show that in the square-law formulation one obtains

$$I_D = \frac{2\pi}{\ln(r_2/r_1)} \bar{\mu}_n C_o \left[(V_G - V_T)V_D - \frac{V_D^2}{2} \right]$$

for below pinch-off operation of a MOSFET with circular geometry. To derive the above result use cylindrical coordinates (r, θ, z) and appropriately modify Eqs. (3.6) through (3.15).

(b) Setting $r_2 = r_1 + L$ and $Z = 2\pi r_1$ show that the part (a) result reduces to the linear geometry result, Eq. (3.15), in the limit where $L/r_1 \ll 1$.

3.10 Making free use of the square-law entries in Table 3.1, ignoring the variation of $\bar{\mu}_n$ with V_G, and *employing only one set of coordinates per each part of the problem*, draw

(a) $g_d/\sqrt{Z\mu_n C_o/L}$ versus V_G $(0 \leq V_G \leq 5$ V$)$ if $V_T = 1$ V and $V_D = 0$, 1, and 2 V.

(b) $g_d/\sqrt{Z\mu_n C_o/L}$ versus V_D $(0 \leq V_D \leq 5$ V$)$ when $V_G - V_T = 1$, 2, and 3 V.

(c) $g_m/\sqrt{Z\mu_n C_o/L}$ versus V_G $(0 \leq V_G \leq 5$ V$)$ if $V_T = 1$ V and $V_D = 1$, 2, and 3 V.

(d) $g_m/\sqrt{Z\mu_n C_o/L}$ versus V_D $(0 \leq V_D \leq 5$ V$)$ when $V_G - V_T = 1$, 2, and 3 V.

3.11 Compare the J-FET and MOSFET; concisely describe similarities and differences in structure, operation, and analysis.

3.12 With modern computing and plotting programs available for use with microcomputers, it is a relatively straightforward task to construct I_D–V_D characteristics based on any of the theories cited herein. It would be generally informative to complete one or more of the following exercises.

(a)/(b) Construct a program to compute and plot I_D–V_D characteristics based on (a) the square-law theory and (b) the bulk-charge theory. To check your program, run it to obtain results that can be compared with the characteristics shown in Fig. 3.9.

(c)/(d) Utilizing the relationships in Appendix D, construct a program to compute and plot I_D–V_D characteristics based on (c) the charge-sheet theory and (d) the exact-charge theory. To check your program, run it to obtain results that can be compared with the characteristics shown in Fig. 3.10 and/or Fig. 3.11.

4 / Nonideal MOS

The ideal structure provides a convenient vehicle for establishing the basic principles of MOS theory in a clear and uncomplicated fashion. Real MOS device structures, however, are never perfectly ideal. In this chapter we examine well-documented deviations from the ideal that have been encountered in MOS device structures. The effect of a nonideality on device characteristics, its identified or suspected physical origin, and methods implemented to minimize the nonideality are noted. Because of ease of fabrication and functional simplicity, the MOS-capacitor has long been the test structure of choice for probing nonidealities. It is understandable, therefore, that the vast majority of nonideal effects are illustrated using MOS-C $C-V$ data. The description herein likewise relies heavily on the comparison of real and ideal MOS-C $C-V$ characteristics. Nevertheless, any deviation from the ideal has a comparable impact on the MOS transistor. To underscore this fact the chapter concludes with a section dealing exclusively with the MOSFET. We discuss how nonidealities can affect the MOSFET threshold voltage, practical ramifications, and in-use methods for adjusting the threshold voltage.

4.1 METAL–SEMICONDUCTOR WORKFUNCTION DIFFERENCE

The energy band diagrams for the isolated components of an Al–SiO$_2$–(p-type) Si system are drawn roughly to scale in Fig. 4.1(a). Upon examining this figure we see that in a real device the energy difference between the Fermi energy and the vacuum level is unlikely to be the same in the isolated metal and semiconductor components of the system; that is, in contrast to the ideal structure, $\Phi_M \neq \Phi_S = \chi + (E_c - E_F)_\infty$. To correctly describe real systems the ideal theory must be modified to account for this metal–semiconductor workfunction difference.

In working toward the required modification, let us first construct the equilibrium ($V_G = 0$) energy band diagram appropriate for the sample system of Fig. 4.1(a). We begin by conceptually connecting a wire between the outer ends of the metal and semiconductor. The two materials are then brought together in a vacuum until they are a distance x_o apart. The connecting wire facilitates the transfer of charge between the metal

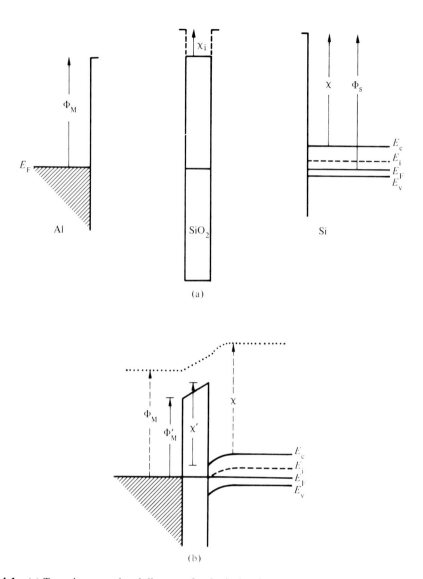

(a)

(b)

Fig. 4.1 (a) To scale energy band diagrams for the isolated components of the Al–SiO$_2$–Si system. (b) Equilibrium ($V_G = 0$) energy band diagram typical of real MOS structures.

and semiconductor and helps maintain the system in an equilibrium state where the respective Fermi levels "line up" as the materials are brought together. With the metal E_F and semiconductor E_F at the same energy, and $\Phi_M \neq \chi + (E_c - E_F)_x$, the vacuum levels in the two materials must be at different energies. Thus an electric field, $\mathscr{E}_{vac}$, develops between the components, with the Si vacuum level above the Al vacuum level given the situation pictured in Fig. 4.1(a), and band bending occurs inside the semiconductor; that is, $K_s\mathscr{E}_s$ must equal $\mathscr{E}_{vac}$. $\mathscr{E}_{vac}$ and the semiconductor band bending increase of course as the components are brought closer and closer together. Once the metal and semiconductor are positioned a distance x_o apart, the insulator is next inserted into the empty space between the other two components. The addition of the insulator simply lowers the effective surface barriers ($\Phi_M \rightarrow \Phi_M - \chi_i = \Phi_M'$ and $\chi \rightarrow \chi - \chi_i = \chi'$) and reduces the electric field in the x_o region ($K_O > 1$). The resulting equilibrium energy band diagram typical of real MOS systems is shown in Fig. 4.1(b).

The point to be derived from the preceding argument and Fig. 4.1(b) is that the workfunction difference modifies the relationship between the semiconductor surface potential and the applied gate voltage. Specifically, setting $V_G = 0$ does not give rise to flat-band conditions inside the semiconductor. Like in a pn junction or MS diode, there is a built-in potential. The precise value of the built-in potential, V_{bi}, can be determined by equating the energies from the Fermi-level to the top of the band diagram as viewed from the two sides of the insulator in Fig. 4.1(b). One obtains

$$\underbrace{\Phi_M' + q\Delta\phi_{ox}'}_{\text{metal side}} = \underbrace{(E_c - E_F)_\infty - q\phi_S + \chi'}_{\text{semiconductor side}} \tag{4.1}$$

Thus, taking the metal to be the zero-potential reference point (the usual procedure in defining built-in potentials), we find

$$V_{bi} = -(\phi_S + \Delta\phi_{ox}) = \phi_{MS} \tag{4.2}$$

where

$$\phi_{MS} \equiv \frac{1}{q}(\Phi_M - \Phi_S) = \frac{1}{q}[\Phi_M' - \chi' - (E_c - E_F)_\infty] \tag{4.3}$$

Perhaps the result here should have been intuitively obvious: The built-in potential inside a $\Phi_M \neq \Phi_S$, but otherwise ideal, MOS structure is just the metal–semiconductor workfunction difference expressed in volts.

In dealing with any nonideality a major concern is the effect of the nonideality on device characteristics. Generally speaking, one would like to know how the given nonideality perturbs the ideal-device characteristics. To illustrate the general determination procedure and to specifically ascertain the effect of a $\phi_{MS} \neq 0$, let us suppose Fig. 4.1(b) is the energy band diagram for an MOS-C. Also let the broken-line curve in Fig. 4.2 be the expected form of the high-frequency $C-V$ characteristic exhibited by

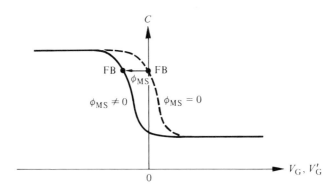

Fig. 4.2   Effect of a $\phi_{MS} \neq 0$ on the MOS-C high-frequency $C-V$ characteristic.

an ideal version of this p-bulk MOS-C. Flat band for the ideal device occurs, of course, at a gate bias of zero volts. On the other hand, from a cursory inspection of Fig. 4.1(b) one infers that a negative bias must be applied to the nonideal device to achieve flat band conditions. In fact, a gate voltage $V_G = \phi_{MS}$ (where $\phi_{MS} < 0$ for the given device) must be applied to offset the built-in voltage and achieve a $\phi_s = 0$. Since both devices will exhibit the same capacitance under flat-band conditions, we conclude the flat-band point for the real device will be displaced laterally ϕ_{MS} volts along the voltage axis.

As it turns out, we can begin the argument just presented at any point along the ideal-device characteristic. There is a one-to-one correspondence between the degree of band bending or ϕ_s and the observed capacitance. Thus, regardless of the reference point along the ideal-device $C-V$ characteristic, one must always apply an added ϕ_{MS} volts to the gate of the real device to achieve the same degree of band bending and hence observe the same capacitance. In other words, as pictured in Fig. 4.2, the entire real-device $C-V$ characteristic will be shifted ϕ_{MS} volts along the voltage axis relative to the ideal-device characteristic.

In the preceding discussion the effect of the $\phi_{MS} \neq 0$ nonideality was described in graphical terms. Alternatively, one can generate a mathematical expression for the voltage shift, ΔV_G, between the ideal and real $C-V$ curves. If V_G' is the voltage applied to the gate of the ideal device to achieve a given capacitance, and V_G the real-device gate voltage required to achieve the same capacitance, then, simply converting the $C-V$ curve discussion into mathematical terms,

$$\Delta V_G = (V_G - V_G')\Big|_{\substack{\text{same } \phi_S \\ \text{(or same } C)}} = \phi_{MS} \tag{4.4}$$

It should be interjected that it is common practice to use V'_G for the gate voltage when referring to the ideal structure. For presentation clarity, we have herein limited the use of V'_G to this chapter where simultaneous reference is made to real and ideal devices.

The actual $\Delta V_G = \phi_{MS}$ value for a given MOS structure is routinely computed from Eq. (4.3) using the $\Phi'_M - \chi'$ appropriate for the system and the $(E_c - E_F)_x$ deduced from a knowledge of the doping concentration inside the semiconductor. The ϕ_{MS} ($T = 300$ K) for the commercially important n^+ poly-Si-gate and Al-gate systems are graphed as a function of doping in Fig. 4.3. The experimentally determined $\Phi'_M - \chi'$ values for a number of other metal–silicon combinations are listed in Table 4.1. Note from Fig. 4.3 and the $\Phi'_M - \chi'$ values listed in Table 4.1 that ϕ_{MS} is more often than not a negative quantity, especially for p-type devices, and is typically quite small — on the order of one volt or less.

SEE EXERCISE 4.1 — APPENDIX A

4.2   OXIDE CHARGES

4.2.1   General Information

As might be inferred from the comments at the end of Section 4.1, $\phi_{MS} \neq 0$ is a relatively minor nonideality. The voltage shift associated with $\phi_{MS} \neq 0$ is small, totally predictable, and incapable of causing device instabilities. Oxide charge, on the other hand, can give rise to far more significant effects, including large voltage shifts and instabilities. Through extensive research a number of distinct charge centers have been identified actually within the oxide or at the Si–SiO$_2$ interface. The nature and position of the oxide charges are summarized in Fig. 4.4.

To establish the general effect of oxide charges, let us postulate the existence of a charge distribution, $\rho_{ox}(x)$, that varies in an arbitrary manner across the width of the oxide layer. Note from the Fig. 4.5 visualization of the charge distribution that, for convenience in this particular analysis, *the origin of the x-coordinate has been relocated at the metal–oxide interface*. With the addition of the charge centers, a portion of the V_G–ϕ_S derivation presented in Subsection 2.3.2 is no longer valid and must be revised. Specifically, in place of Eqs. (2.19) to (2.21), one has, respectively,

$$\frac{d\mathscr{E}_{ox}}{dx} = \frac{\rho_{ox}(x)}{K_O\varepsilon_0} \tag{4.5}$$

$$\mathscr{E}_{ox}(x) = -\frac{d\phi_{ox}}{dx} = \mathscr{E}_{ox}(x_o) - \frac{1}{K_O\varepsilon_0}\int_x^{x_o}\rho_{ox}(x')\,dx' \tag{4.6}$$

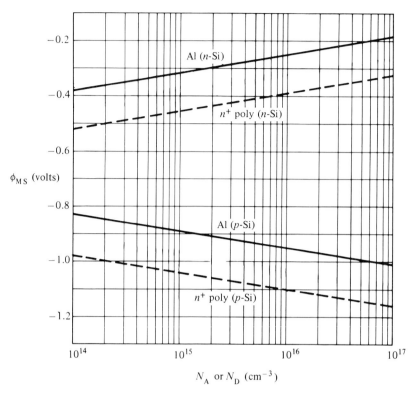

Fig. 4.3 Workfunction difference as a function of the n- and p-type dopant concentration in n^+ poly-Si-gate and Al-gate SiO$_2$–Si structures. ($T = 300$ K. $\Phi'_M - \chi' = -0.18$ eV for the n^+ poly-Si-gate structure; $\Phi'_M - \chi' = 0.03$ eV for the Al-gate structure.)

Table 4.1 Barrier Height Differences in Selected Metal–SiO$_2$–Si Structures.

Metal gate material	$\Phi_M - \chi = \Phi'_M - \chi'$ (eV)
Ag	0.73
Au	0.82
Cr	−0.06
Cu	0.63
Mg	−1.05
Sn	−0.83

Data in the table is taken from S. Kar, *Solid-State Electronics*, **18**, 169 (1975).

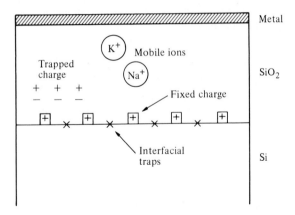

Fig. 4.4 Nature and location of charge centers in thermally grown SiO_2–Si structures. [Adapted from B. E. Deal, *IEEE Trans. on Electron Devices*, **ED-27**, 606, © 1980 IEEE.]

and

$$\Delta\phi_{ox} = x_o \mathscr{E}_{ox}(x_o) - \frac{1}{K_O \varepsilon_0} \int_0^{x_o} \int_x^{x_o} \rho_{ox}(x') \, dx' \, dx \qquad (4.7)$$

The double integral in Eq. (4.7) can be reduced to a single integral employing integration by parts. Moreover, $\mathscr{E}_{ox}(x_o) = K_s \mathscr{E}_s / K_O$ if a plane of charge [other than one possi-

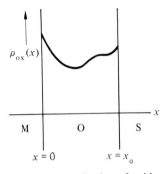

Fig. 4.5 Arbitrary distribution of oxide charges.

bly included in $\rho_{ox}(x)$] is excluded from the oxide–semiconductor interface. Performing the indicated modifications yields

$$\Delta\phi_{ox} = \frac{K_S}{K_O}x_o\mathscr{E}_S - \frac{1}{K_O\varepsilon_0}\int_0^{x_o} x\rho_{ox}(x)\,dx \tag{4.8}$$

Since $V_G = \phi_S + \Delta\phi_{ox}$, for a structure that has charge centers in the oxide but is otherwise ideal, one obtains

$$V_G = \phi_S + \frac{K_S}{K_O}x_o\mathscr{E}_S - \frac{1}{K_O\varepsilon_0}\int_0^{x_o} x\rho_{ox}(x)\,dx \tag{4.9}$$

However, for an ideal device

$$V_G' = \phi_S + \frac{K_S}{K_O}x_o\mathscr{E}_S \tag{4.10}$$

Thus

$$\boxed{\Delta V_G\left(\begin{array}{c}\text{oxide}\\\text{charges}\end{array}\right) = (V_G - V_G')|_{\text{same }\phi_S} = -\frac{1}{K_O\varepsilon_0}\int_0^{x_o} x\rho_{ox}(x)\,dx} \tag{4.11}$$

As emphasized in the development, the voltage translation specified by Eq. (4.11) is valid for an arbitrary charge distribution and is added to the Eq. (4.4) voltage translation due to ϕ_{MS}. In the following subsections we systematically review known information about the various types of charge centers and examine their specific effect on MOS device characteristics.

4.2.2 Mobile Ions

The most perplexing and serious problem encountered in the development of MOS devices can be described as follows: First, the as-fabricated early (c. 1960) devices exhibited C–V characteristics that were sometimes shifted negatively by *tens* of volts with respect to the theoretical characteristics. Second, when subjected to bias-temperature (BT) stressing, a common reliability-testing procedure where a device is heated under bias to accelerate device-degrading processes, the MOS structures displayed a severe instability. The negative shift in the characteristics was increased additional tens of volts after the device was biased positively and heated up to 150 °C or so. Negative bias-temperature stressing had the reverse effect; the $C-V$ curve measured at room temperature after stressing shifted positively or toward the theoretical curve. In extreme cases the instability could even be observed by simply biasing the device at room temperature. One might sweep the $C-V$ characteristics for a given device, go out to lunch leaving the device positively biased, and return to repeat the $C-V$ measurement only to find the characteristics had shifted a volt or so toward negative biases. Note that the characteristics were always shifted in the direction opposite to the applied gate polarity and that the observed curves were always to the negative side of the theoretical curves. The nature and extent of the problem is nicely summarized in Fig. 4.6.

From a practical standpoint, the nonideality causing the as-fabricated translation and instability of the MOS device characteristics had to be identified and eliminated. A device whose effective operating point uncontrollably changes as a function of time is fairly useless. It is now well established that the large as-fabricated shifting and the related instability can be traced to mobile ions inside the oxide, principally Na^+.

If $\rho_{ion}(x)$ is taken to be the ionic charge distribution with

$$Q_M \equiv \int_0^{x_0} \rho_{ion}(x)\, dx \tag{4.12}$$

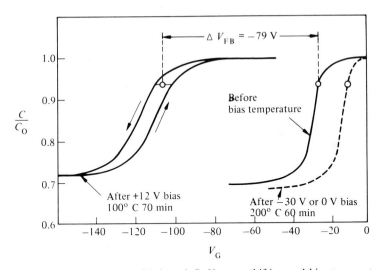

Fig. 4.6 Illustration of the large as-fabricated $C-V$ curve shifting and bias-temperature instability observed with early MOS devices. All $C-V$ curves were taken at room temperature; $x_0 = 0.68\ \mu m$. The arrows adjacent to the after $+BT$ curves indicate the direction of the voltage sweep. [From D. R. Kerr et al., *IBM J. Res. & Dev.*, **8**, 376 (1964). ©1964 by International Business Machines Corporation. Reprinted with permission.]

being the total ionic charge within the oxide per unit area of the gate, then it clearly follows from Eq. (4.11) that

$$\Delta V_G \left(\frac{\text{mobile}}{\text{ions}} \right) = -\frac{1}{K_O \varepsilon_0} \int_0^{x_0} x \rho_{\text{ion}}(x) \, dx \tag{4.13}$$

Note from Eq. (4.13) that positive ions in the oxide would give rise to a negative shift in the $C-V$ characteristics as observed experimentally, while negative ions would give rise to a positive shift in disagreement with experimental observations. Furthermore, because the integrand in Eq. (4.13) varies as $x\rho_{\text{ion}}(x)$, ΔV_G is *sensitive to the exact position of the ions in the oxide*. If, for example, the same Q_M charge per unit gate area is positioned (a) near the metal and (b) near the semiconductor as shown in Fig. 4.7(a),

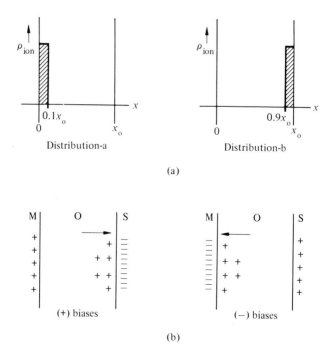

Fig. 4.7 (a) Two hypothetical ionic charge distributions involving the same total number of ions situated near the metal (distribution-a) and near the semiconductor (distribution-b). (b) Expected motion of positive mobile ions within the oxide under $(+)$ and $(-)$ bias-temperature stressing.

one computes $\Delta V_G(a) = -(0.05)Q_M/C_o$ and $\Delta V_G(b) = -(0.95)Q_M/C_o$, where $C_o = K_O\varepsilon_0/x_o$. For the cited example the shift is predicted to be some 19 times larger when the ions are located near the oxide–semiconductor interface! Indeed, based on the preceding observations, it is reasonable to speculate that a large as-fabricated negative shift in the measured C–V characteristics and the attendant instability is caused by *positive* ions in the oxide that *move around or redistribute* under bias-temperature stressing. The required ion movement, away from the metal for $+$BT stressing and toward the metal for $-$BT stressing, is in fact consistent with the direction of ion motion grossly expected from the repulsive/attractive action of other charges within the structure [see Fig. 4.7(b)].

Actual verification of the mobile ion model and identification of the culprit (the ionic species) rivals some of the best courtroom dramas. The suspects were first indicted because of their past history and their accessibility to the scene of the crime. Long before the fabrication of the first MOS device, as far back as 1888, researchers had demonstrated that Na^+, Li^+, and K^+ ions could move through quartz, crystalline SiO_2, at temperatures below 250° C. Furthermore, alkali ions, especially sodium ions, were abundant in chemical reagents, in glass apparatus, on the hands of laboratory personnel, and in the tungsten evaporation boats used in forming the metallic gate. With the suspect identified, great care was taken to avoid alkali ion contamination in the formation of the MOS structure. The net result was devices that showed essentially no change in their C–V characteristics after they were subjected to either positive or negative biases for many hours at temperatures up to 200° C. Next, other carefully processed devices were purposely contaminated by rinsing the oxidized Si wafers in a dilute solution of NaCl (or LiCl) prior to metallization. As expected, the purposely contaminated devices exhibited severe instabilities under bias-temperature stressing. Finally, sodium was positively identified in the oxides of normally fabricated devices (no intentional contamination) through the use of the neutron activation technique; that is, the oxides were bombarded with a sufficient number of neutrons to create a radioactive species of sodium. The analysis of the resultant radioactivity directly confirmed the presence of sodium within the oxide.

Although care to eliminate alkali-ion contamination throughout the fabrication process did lead to stable MOS devices, MOSFET manufacturers encountered difficulties in attaining and maintaining the required degree of quality control in production-line facilities. Thus, in addition to alkali-ion reduction efforts, special fabrication procedures were developed and implemented to minimize the effects of residual alkali-ion contamination. Two different procedures found widespread usage; namely, phosphorus stabilization and chlorine neutralization.

In phosphorus stabilization the oxidized Si wafer is simply placed in a phosphorus diffusion furnace for a short period of time. During the diffusion, as illustrated in Fig. 4.8(a), phosphorus enters the outer portion of the SiO_2 film and becomes incorporated into the bonding structure, thereby forming a new thin layer referred to as a phosphosilicate glass. At the diffusion temperature the sodium ions are extremely mobile and invariably wander into the phosphorus-laden region of the oxide. Once in the

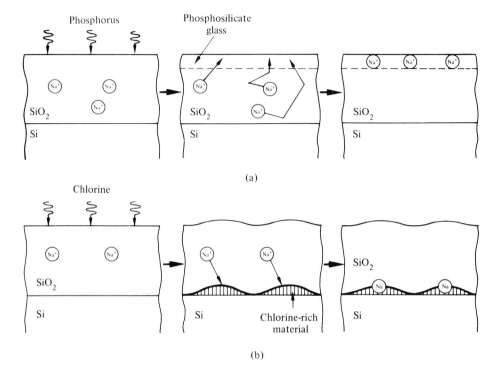

Fig. 4.8 Pictoral description of MOS stabilization procedures: (a) phosphorus stabilization; (b) chlorine neutralization.

phosphosilicate glass the ions become trapped and stay trapped when the system is cooled to room temperature. In this way the alkali ions are "gettered" or drawn out of the major portion of the oxide, are positioned near the outer interface where they give rise to the least amount of as-fabricated $C-V$ curve shifting, and are held firmly in place during normal operating conditions. The phosphosilicate glass layer, it should be noted, also blocks any subsequent contamination associated with the gate metallization or other poststabilization processing steps.

Chlorine neutralization involves a totally different approach. A small amount of chlorine in the form of HCl, Cl_2, or trichloroethane is introduced into the furnace ambient during the growth of the SiO_2 layer. As pictured in Fig. 4.8(b), the chlorine enters the oxide and reacts to form a new material, believed to be a chlorosiloxane, located at the oxide–silicon interface. The new material occupies pancake-shaped regions approximately 1 μm across and several hundred angstroms thick. Stabilization occurs when the ionic sodium migrates into the vicinity of the oxide–silicon interface, becomes trapped, and is *neutralized* by the chlorine in the chlorine-rich material near the interface. Neutral sodium has no effect, of course, on the MOS device characteristics.

The shrinking size of MOS device dimensions, calling for oxide thickness of 0.025 μm = 250 Å or less, limits the use of the cited stabilization procedures in present-day structures. Because of a potential polarization problem, the phosphosilicate glass can be only a small fraction of the overall oxide thickness. With the oxide shrunk to 250 Å or less, the gettering volume becomes small and difficult to control. Likewise, the surface unevenness [see Fig. 4.8(b)] associated with effective chlorine neutralization can be ignored if the oxide is 0.1 μm or thicker. However, the variation in thickness can no longer be tolerated when it becomes comparable to the average oxide thickness. Fortunately, improvements in the purity of fabrication materials (chemicals, gases, etc.) and upgraded processing procedures now permit the direct fabrication of stable MOS devices. It is common practice, nevertheless, to closely monitor furnace tubes and the processing in general to detect the onset of ionic contamination. As an outgrowth of the chlorine neutralization procedure, chlorine has come to be widely employed in the preoxidation cleaning of furnace tubes. Also, a phosphosilicate glass layer is routinely deposited via chemical vapor techniques to form an outer protective coating on ICs. The layer helps to minimize ionic contamination subsequent to device fabrication.

4.2.3 The Fixed Charge

The gross perturbation associated with mobile ions in the oxide tended to obscure or cover up the effects of other deviations from the ideal. Indeed, with the successful elimination of the mobile-ion problem, it became possible to perform a more exacting examination of the device characteristics. The results were rather intriguing. Even in structures free of mobile ions, and after correcting for $\phi_{MS} \neq 0$, the observed $C-V$ characteristics were still translated up to a few volts toward *negative* biases relative to the theoretical characteristics. The possibility of mobile-ion contamination had been eliminated because the structures were stable under bias-temperature stressing. Moreover, for a given set of fabrication conditions observed ΔV_G was *completely reproducible*. Confirming data was obtained from devices independently fabricated by a number of workers at different locations. Subsequent testing (by etching the oxide away in small steps and through photomeasurements) revealed the unexplained ΔV_G shift was caused by a charge residing within the oxide very close to the oxide–semiconductor interface. Because this quasi-interfacial charge was reproducibly fabricated into the structure and was fixed in position under bias-temperature stressing, the nonideality became known as the "built-in" or "fixed" oxide charge.

In modeling the quantitative effect of the fixed charge on the $C-V$ characteristics, it is typically assumed the charge is located immediately adjacent to the oxide–semiconductor interface. Under this assumption one can write

$$\rho_{ox}(x) = Q_F \delta(x_o) \tag{4.14}$$

where $\delta(x_o)$ is a delta-function positioned at the oxide–semiconductor interface and Q_F is the fixed oxide charge per unit area of the gate. Substituting the Eq. (4.14) charge "distribution" into Eq. (4.11) and simplifying yields

$$\Delta V_G\left(\begin{array}{c}\text{fixed}\\\text{charge}\end{array}\right) = -\frac{Q_F}{C_o} \tag{4.15}$$

From Eq. (4.15) it is obvious that, like the mobile-ion charge, the fixed-oxide charge must be *positive* to account for the negative ΔV_G's observed experimentally. Other relevant information about the fixed-oxide charge can be summarized as follows:

1. The fixed charge is independent of the oxide thickness, the semiconductor doping concentration, and the semiconductor doping type (n or p).

2. The fixed charge varies as a function of the Si surface orientation; Q_F is largest on {111} surfaces, smallest on {100} surfaces, and the ratio of the fixed charge on the two surfaces is approximately $3:1$.

3. Q_F is a strong function of the oxidation conditions such as the oxidizing ambient and furnace temperature. As displayed in Fig. 4.9, the fixed charge decreases more or less linearly with increasing oxidation temperatures. It should be emphasized, however, that only the *terminal* oxidation conditions are important. If, for example, a Si wafer is first oxidized in water vapor at $1000°$ C for 1 h, and then exposed to a dry O_2 ambient at $1200°$ C for a sufficiently long time to achieve a steady state condition (~ 5 min), the Q_F value will reflect only the dry oxidation process at $1200°$ C.

4. Annealing (that is, heating) of an oxidized Si wafer in an Ar or N_2 atmosphere for a time sufficient to achieve a steady state condition reduces Q_F to the value observed for the dry oxidations at $1200°$ C. In other words, regardless of the oxidation conditions, the fixed charge can always be reduced to a minimum by annealing in an inert atmosphere.

The preceding experimental facts all provide clues to the physical origin of the fixed-oxide charge. For one, although doping impurities from the semiconductor diffuse into the oxide during the high-temperature oxidation process, the fixed charge was noted to be independent of the semiconductor doping concentration and doping type. The existence of ionized doping impurities within the oxide can therefore be eliminated as a possible source of Q_F. Second, the combination of the interfacial positioning of the fixed charge, the Si-surface orientation dependence, and the sensitivity of Q_F to the terminal oxidation conditions suggests that the fixed charge is intimately related to the oxidizing reaction at the Si–SiO$_2$ interface. In this regard, it should be understood that during the thermal formation of the SiO$_2$ layer, the oxidizing species diffuses through the oxide and reacts at the Si–SiO$_2$ interface to form more SiO$_2$. Thus, the last oxide

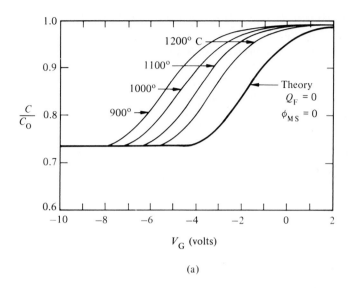

(a)

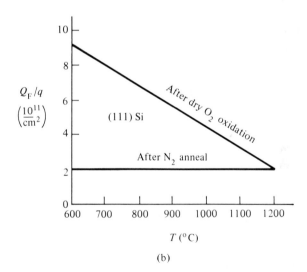

(b)

Fig. 4.9 Effect of the oxidation temperature and annealing on the fixed charge in MOS structures. (a) Measured $C-V$ characteristics after dry O_2 oxidations at various temperatures [x_o = 0.2 μm, N_D = 1.4 × 10^{16}/cm³, (111) Si surface orientation]. (b) Fixed charge concentrations — the so-called *oxidation triangle* specifying the expected Q_F/q after dry O_2 oxidation and after inert ambient annealing. [From B. E. Deal et al., *J. Electrochem. Soc.*, **114**, 266 (1967). Reprinted by permission of the publisher, The Electrochemical Society, Inc.]

formed, the portion of the oxide controlled by the terminal oxidation conditions, lies closest to the Si–SiO$_2$ interface and contains the fixed-oxide charge. From considerations such as these it has been postulated that the fixed-oxide charge is due to *excess ionic silicon* that has broken away from the silicon proper and is waiting to react in the vicinity of the Si–SiO$_2$ interface when the oxidation process is abruptly terminated. The monolayer of oxide adjacent to the Si surface has in fact been experimentally determined to be $x < 2$ SiO$_x$, which is consistent with the excess-Si hypothesis. Annealing in an inert atmosphere, a standard procedure for minimizing the fixed-oxide charge, apparently reduces the excess reaction components and thereby lowers Q_F.

SEE EXERCISE 4.2 — APPENDIX A

4.2.4 Interfacial Traps

Judged in terms of their wide-ranging and degrading effect on the operational behavior of MIS devices, insulator–semiconductor interfacial traps must be considered the most important nonideality encountered in MIS structures. A common manifestation of a nonnegligible interfacial trap concentration within an MOS-C is the distorted or spread out nature of the C–V characteristics. This is nicely illustrated in Fig. 4.10, which dis-

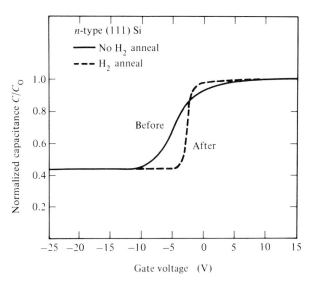

Fig. 4.10 C–V characteristics derived from the same MOS-C before (——) and after (———) minimizing the number of Si–SiO$_2$ interfacial traps inside the structure. [From R. R. Razouk and B. E. Deal, *J. Electrochem. Soc.*, **126**, 1573 (1979). Reprinted by permission of the publisher, The Electrochemical Society, Inc.]

plays two $C-V$ curves derived from the same device before and after minimizing the number of Si–SiO$_2$ interfacial traps inside the structure.

From prior volumes the reader is familiar with donors, acceptors, and recombination–generation (R–G) centers, which introduce localized electronic states in the bulk of a semiconductor. Interfacial traps (also referred to as surface states or interface states) are allowed energy states in which electrons are localized in the vicinity of a material's *surface*. All of the bulk centers are found to add levels to the energy band diagram within the forbidden band gap. Donors, acceptors, and R–G centers, respectively, introduce bulk levels near E_c, E_v, and E_i. Analogously, as modeled in Fig. 4.11, interfacial traps introduce energy levels in the forbidden band gap at the Si–SiO$_2$ interface. Note, however, that interface states can, and normally do, introduce levels distributed throughout the band gap. Interface levels can also occur at energies greater than E_c or less than E_v, but such levels are usually obscured by the much larger density of conduction or valence band states.

Figure 4.12 provides some insight into the behavior and significance of the levels. When an n-bulk MOS-C is biased into inversion as shown in Fig. 4.12(a), the Fermi level at the surface lies close to E_v. For the given situation essentially all of the interfacial traps will be empty because, to a first-order approximation, all energy levels above E_F are empty and all energy levels below E_F are filled. Moreover, if the states are assumed to be donorlike in nature (that is, positively charged when empty and neutral when filled with an electron), the net charge per unit area associated with the interfacial traps, Q_{IT}, will be positive. Changing the gate bias to achieve depletion conditions [Fig. 4.12(b)] positions the Fermi level somewhere near the middle of the band gap at the surface. Since the interface levels always remain fixed in energy relative to E_c and E_v at the surface, depletion biasing obviously draws electrons into the lower interface state levels and Q_{IT} reflects the added negative charge: Q_{IT} (depletion) $< Q_{IT}$ (inversion). Finally, with the MOS-C accumulation biased [Fig. 4.12(c)], electrons fill most of the interfacial traps and Q_{IT} approaches its minimum value. The point is that the interfacial traps charge and discharge as a function of bias, thereby affecting the charge distribu-

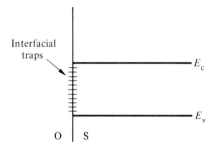

Fig. 4.11 Electrical modeling of interfacial traps as allowed electronic levels localized in space at the oxide–semiconductor interface.

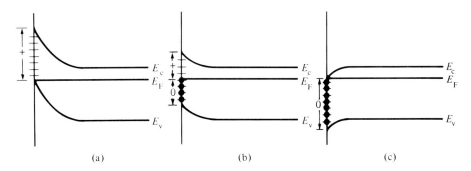

Fig. 4.12 Filling of the interface levels under (a) inversion, (b) depletion, and (c) accumulation biasing in an *n*-type device. The charge state exhibited by donorlike interfacial traps ["+" (plus) or "0" (neutral)] is noted to the left of the respective diagrams.

tion inside the device, the V_G–ϕ_S relationship, and the device characteristics in an understandable but somewhat complex manner.

The gross effect of interfacial traps on the V_G–ϕ_S relationship is actually quite easy to establish. Since Q_{IT}, like Q_F, is located right at the Si–SiO$_2$ interface, we can write by analogy with the fixed charge result

$$\Delta V_G \left(\begin{matrix} \text{interfacial} \\ \text{traps} \end{matrix} \right) = - \frac{Q_{IT}(\phi_S)}{C_o} \qquad (4.16)$$

As emphasized in Eq. (4.16), the result here differs from the fixed-charge result in the Q_{IT} varies with ϕ_S, while Q_F is a constant independent of ϕ_S.

Combined with the earlier considerations on the filling of interface levels, Eq. (4.16) helps to explain the form of the C–V characteristics presented in Fig. 4.10. Assuming donorlike interfacial traps, Q_{IT} takes on its largest positive value under inversion conditions and gives rise to a moderately large negative shift in the C–V characteristics. In progressing through depletion toward accumulation, Q_{IT} decreases, and the translation in the C–V curve likewise decreases as observed experimentally. Once in accumulation ΔV_G should continue to decrease and still remain negative according to the Fig. 4.12 model. The Fig. 4.10 data, on the other hand, exhibits an increasingly *positive* shift in the characteristics with increased accumulation biasing. This discrepancy can be traced to the donorlike assumption. In actual MOS devices the interfacial traps in the upper half of the band gap are believed to be acceptorlike in nature (that is, neutral when empty and negative when filled with an electron). Thus, upon reaching flat band (or roughly flat band), Q_{IT} passes through zero and becomes increasingly negative as more and more upper band gap states are filled with electrons. Qualitatively, then, we can explain the observed characteristics. A complete quantitative description would require a detailed knowledge of the interfacial trap concentration versus energy and additional theoretical considerations to establish an explicit expression for Q_{IT} as a function of ϕ_S.

Although models that detail the electrical behavior of the interfacial traps exist, the *physical origin* of the traps has not been totally clarified. The weight of experimental evidence, however, supports the view that the interfacial traps primarily arise from unsatisfied chemical bonds or so-called "dangling bonds" at the surface of the semiconductor. When the silicon lattice is abruptly terminated along a given plane to form a surface, one of the four surface-atom bonds is left dangling as pictured in Fig. 4.13(a). Logically, the thermal formation of the SiO_2 layer ties up some but not all of the Si-surface bonds. It is remaining dangling bonds that become the interfacial traps [see Fig. 4.13(b)].

To add support to the foregoing physical model, let us perform a simple feasibility calculation. On a (100) surface there are 6.8×10^{14} Si atoms per cm^2. If $1/1000$ of these form interfacial traps and one electronic charge is associated with each trap, the structure would contain a $Q_{IT}/q = 6.8 \times 10^{11}/cm^2$. Choosing an $x_o = 0.2$ μm and substituting into Eq. (4.16) we obtain a ΔV_G (interfacial traps) $= 6.3$ V. Clearly, only a relatively small number of residual dangling bonds can significantly perturb the device characteristics and readily account for observed interfacial trap concentrations.

The overall interfacial trap concentration and the distribution or density of states as a function of energy in the band gap (given the symbol D_{IT} with units of states/cm^2-eV) are extremely sensitive to even minor fabrication details and vary significantly from device to device. Nevertheless, reproducible general trends have been recorded. The interfacial-trap density, like the fixed-oxide charge, is greatest on {111} Si surfaces, smallest on {100} surfaces, and the ratio of midgap states on the two surfaces is approximately $3:1$. After oxidation in a dry O_2 ambient, D_{IT} is relatively high, $\sim 10^{11}$ to 10^{12} states/cm^2-eV at midgap, with the density decreasing for increased oxidation temperatures in a manner also paralleling the fixed-oxide charge. Annealing at high temperatures ($\geq 600°$ C) in an inert ambient, however, does *not* minimize D_{IT}. Rather, as will be described shortly, annealing in the presence of hydrogen at relatively low temperatures ($\leq 500°$ C) minimizes D_{IT}. D_{IT} at midgap after an ideal interface-state anneal is typically $\leq 10^{10}/cm^2$-eV and the distribution of states as a function of energy is of the form sketched in Fig. 4.14. As shown in this figure, the interfacial-trap density is more or less constant over the midgap region and increases rapidly as one approaches the band edges. Lastly, the states near the two band edges are usually about equal in number and opposite in their charging character; that is, states near the conduction and valence bands are believed to be acceptorlike and donorlike in nature, respectively.

The very important annealing of MOS structures to minimize the interfacial-trap concentration is routinely accomplished in one of two ways, namely, through postmetallization annealing or hydrogen (H_2) ambient annealing. In the postmetallization process, which requires a chemically active gate material such as Al or Cr, the metal-

Dangling bond

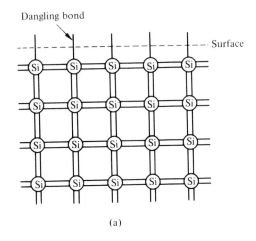

(a)

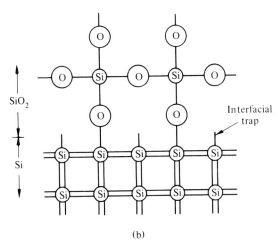

(b)

Fig. 4.13 Physical model for the interfacial traps. (a) "Dangling bonds," which occur when the Si lattice is abruptly terminated along a given plane to form a surface. (b) Postoxidation dangling bonds (relative number greatly exaggerated) that become the interfacial traps. [Part (b) adapted from B. E. Deal, *J. Electrochem. Soc.*, **121**, 198C (1974).]

lized structure is simply placed in a nitrogen ambient at ~450° C for 5 to 10 min. During the formation of MOS structures, minute amounts of water vapor inevitably become absorbed on the SiO_2 surface. At the postmetallization annealing temperature the active gate material reacts with the water vapor on the oxide surface to release a hydrogen species thought to be atomic hydrogen. As pictured in Fig. 4.15, the hydrogen species subsequently migrates through the SiO_2 layer to the Si–SiO_2 interface

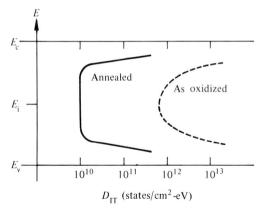

Fig. 4.14 Energy distribution of interface states within the band gap. General form and approximate magnitude of the interfacial-trap density observed before and after an interface-state anneal.

where it attaches itself to a dangling Si bond, thereby making the bond electrically inactive. The hydrogen-ambient process operates on a similar principle, except the hydrogen is supplied directly in the ambient and the structure need not be metallized.

Even though we originally stated that the interfacial trap problem was of paramount importance, it is a challenge to convey the true significance and scope of the problem. Simply stated, if the thermal oxide didn't tie up most of the dangling Si bonds, and if an annealing process were not available for reducing the remaining bonds or interfacial traps to an acceptable level, MOS devices would merely be a laboratory curiosity. Indeed, high interfacial-trap concentrations have severely stunted the development of other insulator–semiconductor systems.

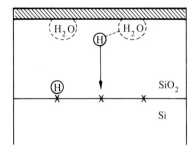

Fig. 4.15 Model for the annihilation of interface states during the postmetallization annealing process. $\textcircled{H}$'s represent the active hydrogen species involved in the process; X's represent interface states.

4.2.5   Induced Charges

Radiation Effects

Radiation damage in solid-state devices has been a major concern of space and military experts since the launch of the first Telstar satellite through the van Allen radiation belt in the early 1960s. Radiation temporarily disabled the Telstar satellite and can have debilitating effects on most solid-state devices and systems. Radiation in the form of X rays, energetic electrons, protons, and heavy ionized particles all have a similar effect on MOS devices. After irradiation, MOS device structures invariably exhibit both an increase in the apparent fixed charge within the oxide and an increase in the interfacial-trap concentration.

The sequence of events leading to radiation-induced damage is summarized in Fig. 4.16. The primary effect directly related to the ionizing radiation is the generation of electron-hole pairs throughout the oxide. A percentage of the generated electrons and holes recombine immediately. The electric field in the oxide operates to separate the surviving carriers, accelerating electrons and holes in opposite directions. Electrons, which have a relatively high mobility in SiO_2, are rapidly (nsec) swept out of the oxide. Holes, on the other hand, tend to be trapped near their point of origin. Over a period of time the holes migrate to the Si–SiO_2 interface (assuming $\mathscr{E}_{ox}$ is positive as shown in Fig. 4.16) where they either recombine with electrons from the silicon or become trapped at deep-level sites. Once trapped in the near vicinity of the interface, the holes mimic the fixed charge thereby giving rise to an apparent increase in Q_F. The process leading to interface-state creation is less well understood. Some of the inter-

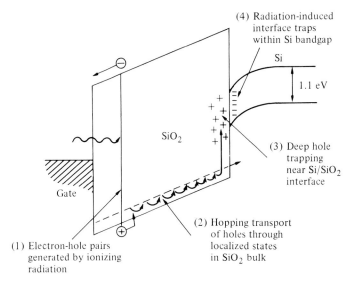

Fig. 4.16   Response to ionizing radiation and the resulting damage in MOS structures. [From J. R. Srour and J. M. McGarrity, *Proc. IEEE*, **76,** 1443, © 1988 IEEE.]

facial traps are created immediately upon exposure to ionizing radiation, while the remainder are created in proportion to the number of holes reaching the Si–SiO$_2$ interface. It has been proposed that the energy released in the deep-level trapping of holes breaks the Si–H bonds associated with deactivated interfacial traps.

Over a period of days to years the trapped-hole charge tends to be slowly reduced at room temperature by the capture of electrons injected into the oxide from the metal or the silicon. Naturally, removal of the trapped-hole charge is greatly accelerated by thermal annealing; a standard interfacial-trap anneal totally removes both the trapped holes and the induced interfacial traps. It should be noted, however, that once subjected to ionizing radiation and recovered through low-temperature annealing, MOS devices exhibit greater sensitivity to subsequent radiation. This has been attributed to the creation of neutral traps within the oxide in addition to the induced charges. Higher-temperature annealing ($T > 600°\,C$) has been found to be partially effective in removing the neutral traps.

Thermal annealing can be readily performed to remove radiation damage that occurs during fabrication.* It is also feasible to increase the ambient temperature of completed devices to 100° C or so to accelerate the removal of trapped holes caused by ionizing radiation. After-the-fact recovery in completed devices, however, is relatively limited. It is preferable to "harden" the devices. The oxide is hardened (i.e., its sensitivity to radiation reduced) by employing empirically established optimum growth conditions such as oxidation temperatures below 1000° C. Other hardening procedures include Al-shielding, which stops the m ority of energetic electrons encountered in space, and increasing the threshold voltage of MOSFETs so they are less sensitive to ΔV_G changes caused by radiation. (The general topic of MOSFET threshold adjustment is discussed in Section 4.3.) Somewhat fortuitously, the reduction in oxide thickness that accompanies reductions in device dimensions is also leading to harder MOS devices. ΔV_G for both the fixed-charge and interfacial traps is proportional to $1/C_o = x_o/K_O\,\varepsilon_0$ and therefore automatically decreases with decreasing x_o. The improvement may also be due, in part, to the smaller hole trapping cross sections at the higher-oxide fields, which exist across the thinner oxides. Projections even suggest hole trapping and the associated interfacial-trap production might actually vanish as the oxide thickness approaches 100 Å. It is envisioned that electrons from the metal and semiconductor can tunnel into all parts of a very thin oxide and rapidly annihilate the trapped holes.

Negative-Bias Instability

Negative-bias instability is a significant perturbation of MOS device characteristics that occurs as a direct result of stressing the structure with a large negative bias at elevated temperatures. Typical stress conditions would be a negative-gate bias sufficient to pro-

---

*Ion implantation, electron-beam evaporation of metals, deposition of special-purpose thin films over the SiO$_2$ layer in a hostile plasma environment (sputtering), electron-beam and X-ray lithography, and a number of other fabrication processes can lead to varying degrees of radiation damage.

duce a field of 2×10^6 V/cm in the oxide and $T > 250°$ C. The instability is characterized by a large *negative* shift along the voltage axis and a distortion of the MOS-C
C–V curve. Similar to ionizing radiation, the stress clearly causes an increase in the
apparent fixed charge within the oxide and an increase in the interfacial-trap concentration. An added peak in D_{IT} near midgap is considered to be a distinctive signature of
the instability. Note that the C–V curve shifting related to the negative-bias instability
is opposite to that caused by alkali-ion contamination.

The exact mechanism causing the negative-bias instability has not been established.
However, because large hole concentrations are adjacent to the oxide under the conditions of the negative stress, it has been proposed that the instability may result from
hole injection from the Si into the oxide and the subsequent trapping of the holes at
deep-level sites near the Si–SiO$_2$ interface. The radiation-induced and stress-induced
damage would then have a related origin. Indeed, it has been found that the negative-
bias instability is enhanced if the MOS structure is first exposed to ionizing radiation.
Sensitivity to the instability can be minimized by annealing the device structure in hydrogen at $800°$ C–$900°$ C prior to gate deposition.

4.2.6 ΔV_G Summary

In the first two sections of this chapter we cited and scrutinized four of the most commonly encountered deviations from the ideal; specifically, the metal–semiconductor
workfunction difference, mobile ions in the oxide, fixed-oxide charge, and interfacial
traps. It was also noted that ionizing radiation and stressing with a large negative bias
at elevated temperatures give rise to additional oxide charges. In both cases there is an
increase in Q_{IT} and the apparent Q_F.

The combined effect of the analyzed nonidealities on the V_G–ϕ_S relationship is described by

$$\Delta V_G = (V_G - V_G')\big|_{\text{same } \phi_S} = \phi_{MS} - \frac{Q_F}{C_o} - \frac{Q_M \gamma_M}{C_o} - \frac{Q_{IT}(\phi_S)}{C_o} \qquad (4.17)$$

where

$$\gamma_M \equiv \int_0^{x_o} \frac{x \rho_{ion}(x)\, dx}{x_o \int_0^{x_o} \rho_{ion}(x)\, dx} \qquad (4.18)$$

In writing down Eq. (4.17) we recast the mobile ion contribution [Eq. (4.13)] to emphasize the similarity between the three terms associated with oxide charges. γ_M is a

unitless quantity representing the centroid of the mobile ion charge in the oxide normalized to the width of the oxide layer; $\gamma_M = 0$ if the mobile ions are all at the metal–oxide interface, while $\gamma_M = 1$ if the mobile ions are all piled up at the Si–SiO$_2$ interface. As a general rule, Q_M, Q_F, and ϕ_{MS} all lead to a negative parallel translation of the C–V characteristics along the voltage axis relative to the ideal theory. ΔV_G due to Q_{IT}, on the other hand, can be either positive or negative, depends on the applied bias, and tends to distort or spread out the characteristics.

From the discussion of nonidealities, it is safe to conclude that real MOS devices are not intrinsically perfect; the devices are in fact intrinsically imperfect. Through an extensive research effort, however, procedures have been developed for minimizing the net effect of nonidealities in MOS structures. Although constant checks must be run to maintain quality control, manufacturers today routinely fabricate near-ideal devices. We should mention that, whereas the described or similar nonidealities apply to all MIS structures, the specific minimization procedures outlined herein apply only to the thermally-grown SiO$_2$-Si system.

> **SEE EXERCISE 4.3—APPENDIX A**

4.3 MOSFET THRESHOLD CONSIDERATIONS

Thus far we have described the effect of nonidealities in terms of MOS-C C–V curve shifting and distortion. The V_T point on the C–V curve, corresponding to the threshold voltage of a MOSFET, is of course directly affected by any ΔV_G displacement. Figure 4.17(a) graphically illustrates the general effect of changes in the threshold voltage on the MOSFET I_D–V_D characteristics. A hypothetical p-channel MOSFET with an ideal device $V'_T = -1$ V was assumed in constructing the figure. The ΔV_G leading to the perturbed characteristics in Fig. 4.17(a) is taken to be caused by ϕ_{MS}, the fixed charged, and/or alkali ions in the oxide. Note that, analogous to the parallel translation of the MOS-C C–V characteristics, the form of the perturbed MOSFET characteristics is unchanged, but larger $|V_G|$ values are needed to achieve comparable I_D current levels. If present in large densities, the interfacial traps can additionally decrease the change in current resulting from a stepped increase in gate voltage. This effect, equivalent to a reduction in the g_m of the transistor, is pictured in Fig. 4.17(b).

While it is true that both the mobile ion and interfacial trap problems were minimized early in MOSFET development, the remaining nonidealities, primarily through their effect on V_T, have had a very large impact on fabrication technology, device design, and modes of operation. In this section we examine the reasons for the impact and methods that have evolved for adjusting the MOSFET threshold.

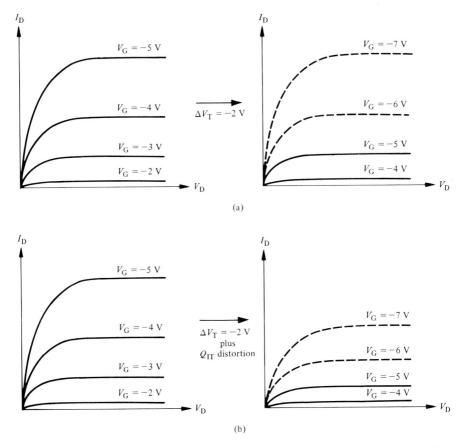

Fig. 4.17 General effect of nonidealities on the MOSFET current-voltage characteristics. The ideal characteristics of a hypothetical p-channel MOSFET with $V_T' = -1$ V are pictured on the left and the perturbed characteristics are shown on the right. (a) Effect of a simple shift in the threshold voltage caused by $\phi_{MS} \neq 0$, the fixed charge, and/or mobile ions. (b) g_m degrading effect of interfacial traps.

4.3.1 V_T Relationships

For discussion purposes it is desirable to establish an expression for the expected threshold voltage, V_T, of a real MOSFET. Let V_T' be the threshold voltage of an ideal version of the given MOSFET. Simply evaluating Eq. (4.17) at $\phi_s = 2\phi_F$ then gives

$$V_T = V_T' + \phi_{MS} - \frac{Q_F}{C_o} - \frac{Q_M \gamma_M}{C_o} - \frac{Q_{IT}(2\phi_F)}{C_o} \tag{4.19}$$

Although Eq. (4.19) can be used directly to compute V_T, it is standard practice to express the threshold shift in terms of the real-device flat-band voltage. Under flat-band conditions $\phi_S = 0$, $V_G' = 0$, and from Eq. (4.17),

$$V_{FB} \equiv V_G|_{\phi_S=0} = \phi_{MS} - \frac{Q_F}{C_o} - \frac{Q_M \gamma_M}{C_o} - \frac{Q_{IT}(0)}{C_o} \qquad (4.20)$$

If Q_{IT} changes little in going from $\phi_S = 0$ to $\phi_S = 2\phi_F$, a reasonably good approximation in well-made devices, the nonideality-related terms in Eq. (4.19) can be replaced by V_{FB} and one obtains

$$V_T = V_T' + V_{FB} \qquad (4.21)$$

where, repeating Eqs. (3.1),

$$V_T' = 2\phi_F \pm \frac{K_S}{K_O} x_o \sqrt{\frac{4qN_B}{K_S \varepsilon_0}(\pm\phi_F)} \qquad \begin{array}{l} (+) \text{ for } n\text{-channel devices} \\ (-) \text{ for } p\text{-channel devices} \\ N_B = N_A \text{ or } N_D \text{ as appropriate} \end{array} \qquad (4.22)$$

4.3.2   Threshold, Terminology, and Technology

As an entry into the discussion, let us perform a simple threshold voltage computation employing relationships developed in Subsection 4.3.1. Suppose the gate material is Al, the Si surface orientation is (111), $T = 300$ K, $x_o = 0.1$ μm, $N_A = 10^{15}/\text{cm}^3$, $Q_F/q = 2 \times 10^{11}/\text{cm}^2$, $Q_M = 0$, and $Q_{IT} = 0$. For the given n-channel device one computes $\phi_{MS} = -0.89$ V, $-Q_F/C_o = -0.93$ V, $V_{FB} = -1.82$ V, $V_T' = 0.99$ V, and $V_T = -0.83$ V. Observe that whereas V_T' is positive, as expected, *nonidealities of a very realistic magnitude cause V_T to be negative.* Since an n-channel device turns on for gate voltages $V_G > V_T$, the device in question is already "on" at a gate bias of zero volts. Actually, negative biases must be applied to turn the device off! For a p-channel device with identical parameters (except, of course, for an N_D doped substrate) one obtains a $V_T' = -0.99$ V, $V_{FB} = -1.24$ V, and $V_T = -2.23$ V. In the p-channel case the considered nonidealities merely increase the negative voltage required to achieve turn-on.

When a MOSFET is "on" at $V_G = 0$ V, the transistor is referred to as a *depletion mode* device; when a MOSFET is "off" at $V_G = 0$ V, it is called an *enhancement mode* device. Routinely fabricated p-channel MOSFETs constructed in the standard configuration are ideally and practically enhancement mode devices. *n*-channel MOSFETs are also ideally enhancement mode devices. However, because nonidealities tend to shift the threshold voltage toward negative biases in the manner indicated in our sample calculation, early *n*-channel MOS transistors were typically of the depletion mode type. Up until approximately 1977 this difference in behavior led to the total dominance of *PMOS* technology over *NMOS* technology; that is, ICs incorporating *p*-channel MOSFETs dominated the commercial marketplace. Subsequently, as explained under the heading of threshold adjustment, NMOS, which is to be preferred because of the greater mobility of electrons compared to holes, benefited from technological innovations widely implemented in the late 1970s and is now incorporated in the majority of newly designed ICs.

While on the topic of the threshold voltage in practical devices, it is relevant to note that the inversion threshold of regions adjacent to the device proper is also of concern. Consider, for example, the unmetallized region between the two *n*-channel MOS transistors pictured in Fig. 4.18(a). If the potential at the unmetallized outer oxide sur-

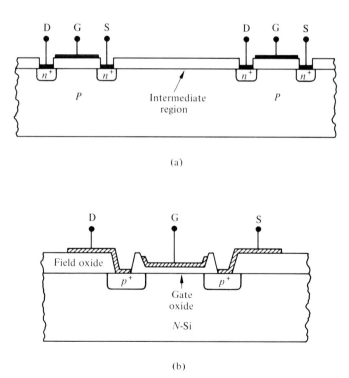

Fig. 4.18   (a) Visualization of the intermediate region between two MOSFETs. (b) Identification of the gate-oxide and field-oxide regions in practical MOSFET structures.

face is assumed to be zero (normally a fairly reasonable assumption) and if the threshold voltage for the n-channel transistors is negative, then the intermediate region between the two transistors will be inverted. In other words, a conducting path, a pseudo-channel, will exist between the transistors. This undesirable condition was another nuisance in early NMOS technology, where, as already noted, nonidealities tended to invert the surface of the semiconductor in the absence of an applied gate bias. Unless special precautions are taken, unwanted pseudo-channels between devices can also arise in both n- and p-channel ICs from the potential applied to the metal overlays supplying the gate and drain biases. To avoid this problem the oxide in the nongated portions of the IC, referred to as the *field-oxide*, is typically much thicker than the *gate-oxide* in the active regions of the structure [see Fig. 4.18(b)]. The idea behind the use of the thicker oxide can be understood by referring to Eqs. (4.20) and (4.22). Both V_{FB} and V_T' contain terms that are proportional to x_o. Thus employing an x_o (field-oxide) $\gg x_o$ (gate-oxide) increases $|V_T|$ in the field-oxide areas relative to the gated areas in PMOS (and modern NMOS) structures. Inversion of the field-oxide regions is thereby avoided at potentials normally required for IC operation.

4.3.3 Threshold Adjustment

Several physical factors affect the threshold voltage and can therefore be used to vary the V_T actually exhibited by a given MOSFET. We have, in fact, already cited the adjustment of V_T through a variation of the oxide thickness. Obviously, the substrate doping can also be varied to increase or decrease the threshold voltage. However, although strongly influencing the observed V_T value, the gate-oxide thickness and substrate doping are predetermined in large part by other design restraints.

Other factors that play a significant role in determining V_T are the substrate-surface orientation and the material used in forming the MOS gate. As first noted in Subsection 4.2.3, the Q_F in MOS devices constructed on (100) surfaces is approximately three times smaller than the Q_F in devices constructed on (111) surfaces. The use of (100) substrates therefore reduces the ΔV_G associated with the fixed oxide charge. The use of a polysilicon instead of an Al gate, on the other hand, changes ϕ_{MS}. Given a polysilicon gate the effective "metal" workfunction becomes

$$``\Phi_M" = \chi_{Si} + (E_c - E_F)_{poly\text{-}Si} \qquad (4.23)$$

and

$$\phi_{MS} = \frac{1}{q}[(E_c - E_F)_{poly\text{-}Si} - (E_c - E_F)_{\infty, crystalline\text{-}Si}] \qquad (4.24)$$

If the calculation in Subsection 4.3.2 is revised assuming a (100) surface orientation ($Q_F/q = 2/3 \times 10^{11}/cm^2$) and a p-type polysilicon gate where $E_F = E_v$, one obtains a

$\phi_{MS} = +0.26$ V, $V_{FB} = -0.05$ V, and $V_T = +0.94$ V. Thus positive NMOS thresh- olds are possible in (100)-oriented structures incorporating p-type polysilicon gates.

Although the foregoing calculation shows positive threshold voltages are possible, the V_T in actual structures may be only nominally positive, or the structure impractical — in forming n-channel MOSFETs, the polysilicon is conveniently doped n-type, the same as the drain and source islands, not p-type as assumed above. Also, a larger threshold voltage may be desired, or one may desire to modify the threshold attainable in a PMOS structure, or tailoring of the threshold for both n- and p-channel devices on the same IC chip may be required. For a number of reasons, it is very desirable to have a flexible threshold adjustment process where V_T can be controlled essentially at will. In modern device processing this is accomplished through the use of *ion implantation*.

The general ion-implantation process is described in Subsection 1.1.4 of Volume II; an entire chapter is devoted to the process in Volume V. To adjust the threshold voltage, a relatively small, precisely controlled number of either boron or phosphorus ions is im- planted into the near-surface region of the semiconductor. When the MOS structure is depletion or inversion biased, the implanted dopant adds to the exposed dopant-ion charge near the oxide–semiconductor interface and thereby translates the V_T exhibited by the structure. The implantation of boron causes a positive shift in the threshold voltage; phosphorus implantation causes a negative voltage shift. For shallow implants the procedure may be viewed to first order as placing an additional "fixed" charge at the oxide–semiconductor interface. If N_I is the number of implanted ions/cm^2 and $Q_I = \pm q N_I$ is the implant-related donor ($+$) or acceptor ($-$) charge/cm^2 at the oxide– semiconductor interface, then, by analogy with the fixed charge analysis

$$\Delta V_G \left(\begin{array}{c} \text{implanted} \\ \text{ions} \end{array} \right) = -\frac{Q_I}{C_o} \tag{4.25}$$

Assuming, for example, an $N_I = 5 \times 10^{11}$ boron ions/cm^2 and an $x_o = 0.1$ μm, one computes a threshold adjustment of $+2.32$ V.

4.3.4 Back Biasing

Reverse biasing the back contact or bulk of a MOS transistor relative to the source is another method that has been employed to adjust the threshold potential. This electri- cal method of adjustment, which predates ion implantation, makes use of the so-called *body effect* or substrate-bias effect.

To explain the effect let us consider the n-channel MOSFET shown in Fig. 4.19(a). If the back-to-source potential difference (V_{BS}) is zero, inversion occurs of course when the voltage drop across the semiconductor (ϕ_S) equals $2\phi_F$ as pictured in Fig. 4.19(b). If $V_{BS} < 0$, the semiconductor still attempts to invert when ϕ_S reaches $2\phi_F$. However, with $V_{BS} < 0$ any inversion-layer carriers that do appear at the semiconductor surface migrate laterally into the source and drain because these regions are at a lower poten-

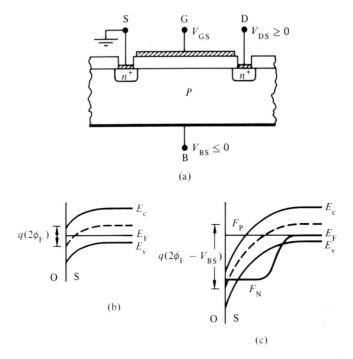

Fig. 4.19 The back-biased MOSFET. (a) Cross-sectional view indicating the double subscripted voltage variables used in the analysis. Also shown are the semiconductor energy band diagrams corresponding to the onset of inversion when (b) $V_{BS} = 0$ and (c) $V_{BS} < 0$.

tial. Not until $\phi_S = 2\phi_F - V_{BS}$, as pictured in Fig. 4.19(c), will the surface invert and normal transistor action begin. In essence, back biasing changes the inversion point in the semiconductor from $2\phi_F$ to $2\phi_F - V_{BS}$. The ideal device threshold potential given by Eq. (4.22) is in turn modified to

$$V'_{GB\,|\,\text{at threshold}} = 2\phi_F - V_{BS} \pm \frac{K_S}{K_O}x_o\sqrt{\frac{2qN_B}{K_S\varepsilon_0}(\pm 2\phi_F \mp V_{BS})}$$

$$(+),\ V_{BS} < 0 \text{ for } n\text{-channel}$$
$$(-),\ V_{BS} > 0 \text{ for } p\text{-channel} \qquad (4.26)$$

or, since $V'_{GB\,|\,\text{at threshold}} = V'_{GS|\text{at threshold}} - V_{BS}$,

$$V'_{GS\,|\,\text{at threshold}} = 2\phi_F \pm \frac{K_S}{K_O}x_o\sqrt{\frac{2qN_B}{K_S\varepsilon_0}(\pm 2\phi_F \mp V_{BS})} \qquad (4.27)$$

Finally, introducing $\Delta V'_T \equiv (V'_{GS|\text{at threshold}} - V'_T)$, we obtain

$$\Delta V'_T = (V'_T - 2\phi_F)\left[\sqrt{1 - \frac{V_{BS}}{2\phi_F}} - 1\right] \quad \begin{array}{l} \phi_F > 0, \ V_{BS} < 0 \text{ for } n\text{-channel} \\ \phi_F < 0, \ V_{BS} > 0 \text{ for } p\text{-channel} \end{array} \quad (4.28)$$

Having established Eq. (4.28), we make the following observations concerning back biasing or the body effect: (1) Back biasing always increases the magnitude of the ideal device threshold voltage. It therefore makes the p-channel threshold of actual devices more negative and the n-channel threshold more positive — it cannot be used to reduce the negative threshold of a p-channel MOSFET. (2) The current–voltage relationships developed in Chapter 3 are still valid when $V_{BS} \neq 0$ provided $2\phi_F \rightarrow 2\phi_F - V_{BS}$, $V_G \rightarrow V_{GS}$, $V_D \rightarrow V_{DS}$, and V_T is interpreted as $V_{GS|\text{at threshold}}$. (3) Care must be exercised in describing back-biased structures to properly identify voltage differences through the use of double-subscripted voltage variables.

4.3.5 Threshold Summary

All nonidealities shift the MOSFET threshold voltage; interfacial traps in addition reduce the low-frequency g_m of the transistor. A two-step procedure is followed in computing the expected threshold voltage of a real MOSFET. The flat-band voltage is first deduced from known information about device nonidealities. Subsequently V_{FB} is added to the threshold voltage of an ideal version of the given MOSFET. A transistor that is normally "off" when $V_G = 0$ is referred to as an enhancement mode device; a depletion mode MOSFET is "on" or conducting when $V_G = 0$. Because of residual nonidealities, real n-channel MOSFETs are typically depletion-mode devices. The attainment of n-channel enhancement mode MOSFETs, and threshold adjustment in general, is now accomplished through the use of ion implantation. Biasing the back contact relative to the sources is also employed to adjust the threshold voltage.

SEE EXERCISE 4.4 — APPENDIX A

PROBLEMS

4.1 Answer the following questions as concisely as possible.

(a) How is the semiconductor "workfunction" defined?

(b) What is involved in subjecting a MOS-C or MOSFET to a bias-temperature stress?

(c) What is believed to be the physical origin of the fixed-oxide charge in MOS structures?

(d) The interfacial-trap density (D_{IT}) is observed to depend on the silicon surface orientation. Describe the observed orientation dependence.

(e) What is the net effect of ionizing radiation on MOS structures?

(f) When performing bias-temperature stress experiments, how does one distinguish between the negative-bias instability and the voltage instability arising from alkali ions?

(g) Under what circumstances would $V_T \neq V_T' + V_{FB}$?

(h) Explain what is meant by the term "depletion mode" transistor.

(i) What is the difference between the "field-oxide" and the "gate-oxide" in MOSFETs?

(j) Precisely what is the "body effect"?

4.2 Consider a polysilicon-gate MOS-C where $E_F - E_c = 0.2$ eV in the heavily doped gate and $E_F - E_c = -0.2$ eV in the nondegenerately doped silicon substrate. Assume the structure to be ideal (other than an obvious $\phi_{MS} \neq 0$) and $\chi'(\text{poly-Si}) = \chi'(\text{crystalline-Si})$.

(a) Sketch the energy band diagram for the polysilicon-gate MOS-C under flat band conditions.

(b) What is the "metal"–semiconductor workfunction difference for the cited polysilicon-gate MOS-C?

(c) Will the given MOS-C be accumulation, depletion, or inversion biased when $V_G = 0$? Explain.

(d) A nonzero fixed-oxide charge typically appears at a Si–SiO_2 interface. What effect would a $Q_F \neq 0$ at the *polysilicon-gate*–SiO_2 interface have upon the $C-V$ characteristics derived from the MOS-C? Explain.

4.3 In modeling the quantitative effect of the fixed charge, it is typically assumed the charge is located immediately adjacent to the oxide–semiconductor interface. Suppose the charge is actually distributed a short distance into the oxide from the Si–SiO_2 interface.

(a) For reference purposes, write down the standard result for $\Delta V_G(\text{fixed charge})$.

(b) Determine the expected ΔV_G shift caused by an equivalent amount of charge distributed in a linearly increasing fashion from zero at a distance Δx from the Si–SiO_2 interface to $2Q_F/\Delta x$ right at the Si–SiO_2 interface.

(c) Compute $\Delta V_G(\text{part b})/\Delta V_G(\text{part a})$ assuming $\Delta x = 10$ Å $= 10^{-7}$cm and $x_o = 0.1$ μm $= 10^{-5}$cm. Repeat for $x_o = 0.01$ μm $= 10^{-6}$cm. Comment on your results.

4.4 (a) An MOS-C is found to possess a uniform distribution of sodium ions in the oxide; that is, $\rho_{ion}(x) = \rho_0 = $ constant for all x in the oxide. Compute the ΔV_G shift resulting from this distribution of ions if $\rho_0/q = 10^{18}/\text{cm}^3$ and $x_o = 0.1$ μm.

(b) After positive bias-temperature (+BT) stressing, the sodium ions of part (a) all pile up immediately adjacent to the oxide–semiconductor interface. Determine the ΔV_G after +BT stressing.

4.5 If interfacial traps are associated with residual "dangling bonds" at the Si surface, and assuming the number of residual "dangling bonds" is proportional to the number of Si surface atoms, which silicon surface plane, (100) or (110), would be expected to exhibit the higher density of interfacial traps? Record all work leading to your answer.

4.6 A rather unusual n-bulk MOS-C is found to have interfacial traps at only one band-gap energy, E_{IT}, located right at midgap (see Fig. P4.6). Assuming a high-frequency C–V measurement and an otherwise ideal MOS-C,

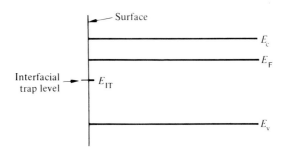

Fig. P4.6

(a) Sketch the general form of the expected MOS-C C–V characteristic if the interface states giving rise to the E_{IT} level are donorlike in nature. (Assume the number of states is sufficiently large to perturb the ideal-device characteristic.)

(b) Repeat part (a) if the interface states are acceptorlike in nature.

(c) Repeat part (a) assuming the interface states are donorlike but the energy level is located very close to the conduction band (say $E_c - E_{IT} = 0.001$ eV).

(d) Repeat part (a) for donorlike interfacial traps that introduce an energy level very close to the valence band (say $E_{IT} - E_v = 0.001$ eV).

Include a few words of explanation, if necessary, to convey your thought process and to prevent a misinterpretation of your sketches. Also briefly discuss your results.

4.7 For demonstration purposes, a university instructor sets out to construct an n-bulk MOS-C with a flat band voltage (V_{FB}) equal to zero. The instructor plans to employ gold as the gate ma-

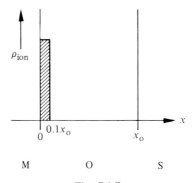

Fig. P4.7

terial and (100) Si wafers with $N_D = 10^{15}/\text{cm}^3$. The fabrication processing is known to yield a $Q_F/q = 6 \times 10^{10}/\text{cm}^2$, $Q_M/q = 2 \times 10^{11}/\text{cm}^2$, and a constant $D_{IT} = 2 \times 10^{10}/\text{cm}^2\text{-eV}$ for all band gap energies. The interfacial traps are acceptorlike in the upper half of the band gap and donorlike in the lower half of the band gap. The device is phosphorus gettered so the alkali ions wind up being distributed in the oxide as shown in Fig. P4.7. Determine the oxide thickness, x_o, the instructor must use to obtain the desired $V_{FB} = 0$. Assume $T = 300$ K.

4.8 An n-channel (p-bulk) MOSFET is ideal except $\phi_{MS} \neq 0$. Indicate how the following modifications to the structure, taken independently, will affect the threshold voltage V_T. Include a few words of explanation in each case.

(a) Ionizing radiation causes an apparent $Q_F \neq 0$.

(b) There is an increase in the substrate doping.

(c) The oxide thickness is decreased.

(d) Boron ions are ion implanted into the near surface region of the silicon.

4.9 The before bias-tempreature stressing I_D–V_D characteristics of a MOSFET are sketched in Fig. P4.9(a). The g_d–V_G ($V_D = 0$) characteristics of the same device before and *after* bias-temperature stressing using a positive bias are shown in Fig. P4.9(b).

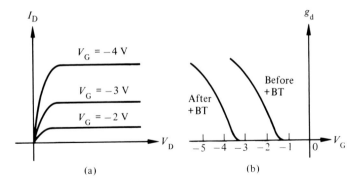

(a)   (b)

Fig. P4.9

(a) What was the probable cause of the shift in the g_d–V_G characteristic after positive bias-temperature stressing?

(b) Sketch the I_D–V_D characteristics *after* positive bias-temperature stressing for gate voltages of $V_G = -2$, -3, and -4 V. Also record your reasoning.

4.10 A MOSFET is fabricated with $\phi_{MS} = -0.46$ V, $Q_F/q = 2 \times 10^{11}/\text{cm}^2$, $Q_M = 0$, $Q_{IT} = 0$, and $Q_I/q = -4 \times 10^{11}/\text{cm}^2$, $x_o = 500$ Å, $A_G = 10^{-3}$ cm^2, and $N_D = 10^{15}/\text{cm}^3$. Q_I is the ion charge implanted immediately adjacent to the Si–SiO$_2$ interface.

(a) Determine V_{FB}.

(b) Determine V_T.

(c) Is the given MOSFET an enhancement mode or depletion mode device? Explain.

4.11 Given an Al-SiO$_2$-Si MOSFET, $T = 300$ K, a Si-substrate doping of $N_A = 5 \times 10^{15}/cm^3$, $x_o = 0.05$ μm, $Q_F/q = 10^{11}/cm^2$, no interfacial traps and no mobile ions in the oxide. Determine the boron ions/cm^2 (N_I) that must be implanted into the structure to achieve a $V_T = 1$ V threshold voltage. (Assume the implanted ions create an added negative charge right at the Si–SiO$_2$ interface.)

5 / Modern FET Structures

To achieve higher operating speeds and increased packing densities, FET-device structures have been subject to greater and greater miniaturization. The decrease in FET-device dimensions in itself can lead to major modifications in the observed device characteristics. Small-dimension effects, also referred to as short-channel effects or small-geometry effects, include, for example, shifts in the threshold voltage and an increase in the subthreshold current. The cited modifications in device behavior are of major importance in practical applications; for example, an accurate prediction of the threshold voltage is needed to determine logic levels, noise margins, speed, and node voltages, while the subthreshold current affects the off-state power dissipation, dynamic logic clock speeds, and memory refresh times. The majority of this chapter is devoted to the description and discussion of small-dimension effects. It should be understood from the outset that small-dimension effects are generally undesirable and are minimized or avoided in commercial structures through the proper scaling of device dimensions or modifications in device design. Relative to device design, the chapter concludes with a brief survey of select implemented and developmental FET structures.

5.1 SMALL-DIMENSION EFFECTS

5.1.1 Introduction

In 1965 the smallest MOSFETs had an $L \sim 1$ mil $= 25$ μm. In 1990, industry-standard versions of MOS device structures boast submicron dimensions. The steady progression toward smaller and smaller FETs is pictured schematically in Fig. 5.1.

The departure from long-channel behavior, which can accompany the noted decrease in device dimensions, is nicely illustrated using observed I_D–V_D characteristics. The *onset* of short-channel effects is heralded by a significant upward slant in the postpinch-off portions of the I_D–V_D curves. *Severe* short-channel effects lead to characteristics of the form reproduced in Fig. 5.2. Not only do the I_D–V_D characteristics fail to saturate, but $I_D \propto V_D^2$ curves are observed for gate voltages below threshold. (The Fig. 5.2 device should be "off" for $V_D > 0$ V!) Another clear manifestation of short-channel effects is

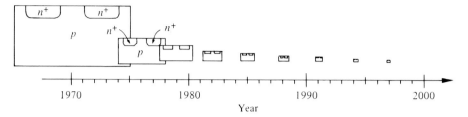

Fig. 5.1 The "shrinking" MOSFET illustrates the historical decrease in minimum feature size of production-line MOS DRAMs (drawn to scale).

provided by the subthreshold-transfer characteristics. A sample long-channel subthreshold characteristic was presented in Fig. 3.11. In long-channel MOSFETs the subthreshold-drain current varies exponentially with V_G and is independent of V_D provided $V_D >$ few kT/q volts. In short-channel devices, on the other hand, the subthreshold-drain current is found to increase systematically and significantly with increasing V_D. Shifts in the observed threshold voltage constitutes the third widely quoted indication of small-dimension effects. As is readily confirmed by referring to the V_T relationship in Subsection 4.3.1, the threshold voltage in a long-channel MOSFET is independent of

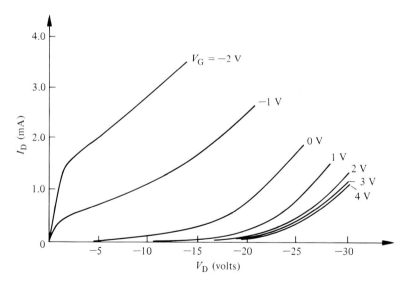

Fig. 5.2 I_D–V_D characteristics of a MOSFET exhibiting severe short-channel effects. [Reprinted with permission from *Solid-State Electronics*, **17**, I. M. Bateman, G. A. Armstrong, and J. A. Magowan, © 1974, Pergamon Press plc.]

the gate length and width. In short-channel devices, V_T becomes a function of the gate dimensions and the applied biases (see Fig. 5.3).

The majority of small-dimension effects in MOSFETs are associated with the reduction in the channel length L. It is therefore reasonable to introduce and specify a minimum channel length, L_{min}, below which significant short-channel effects are expected to occur. Crudely speaking, L_{min} must be greater than the sum of the depletion widths associated with the source and drain junctions. Values in the 0.1 μm to 1 μm range are obviously indicated. As suggested by computer simulations and confirmed by experimental observations, a more precise estimate of L_{min} is given by the empirical relationship*

$$L_{min} = 0.4[r_j x_o (W_S + W_D)^2]^{1/3} \qquad \begin{array}{l} \ldots x_o \text{ in Å; } L_{min}, r_j, \\ W_S, W_D \text{ in } \mu m \end{array} \qquad (5.1)$$

r_j is the source/drain junction depth, x_o the oxide thickness, W_S the depletion width at the source junction, and W_D the depletion width at the drain junction. Note from

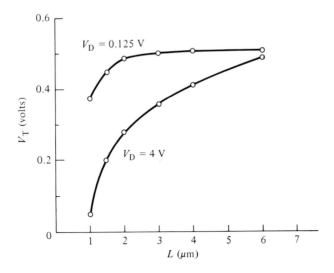

Fig. 5.3 Observed threshold-voltage variation with channel length and applied drain bias in short-channel MOSFETs. $N_A = 8 \times 10^{15}/\text{cm}^3$, $x_o = 0.028$ μm, and $r_j = 1$ μm. [Adapted from W. Fichtner and H. W. Potzl, *Int. J. Electron.*, **46**, 33 (1979).]

---

*J. R. Brews, W. Fichtner, E. H. Nicollian, and S. M. Sze, "Generalized Guide for MOSFET Minimization," *IEEE Electron Device Lett.*, **EDL-1**, 2 (1980).

Eq. (5.1) that L_{min} can be made smaller by reducing the depth of the source/drain islands, by reducing the oxide thickness, and/or by increasing the substrate doping, which in turn causes a decrease in W_S and W_D. All of the above have in fact been employed to help assure long-channel operation of MOSFETs with increasingly scaled-down dimensions.

The causes underlying departures from long-channel behavior fall into one of three general categories. For one, differences between experiment and long-channel theory may simply arise from a breakdown of assumptions used in the long-channel analysis. Second, a reduction in device dimensions automatically leads to an enhancement of certain effects that are known to occur but are negligible in long-channel devices. Lastly, some departures from long-channel behavior arise from totally new phenomena. All three categories are represented in the following consideration of specific-case effects.

5.1.2 Threshold Voltage Modification

Short Channel

In enhancement-mode short-channel devices $|V_T|$ is found to monotonically decrease with decreasing channel length L. Qualitatively, the decrease in threshold voltage can be explained as follows: Before an inversion layer or channel forms beneath the gate, the subgate region must first be depleted ($W \Rightarrow W_T$). In a short-channel device the source and drain assist in depleting the region under the gate; that is, a significant portion of the subgate deletion-region charge is balanced by the charge on the other side of the source and drain pn junctions. Thus less gate charge is required to reach the start of inversion and $|V_T|$ decreases. The smaller L, the greater the percentage of charge balanced by the source/drain pn junctions, and the greater the reduction in $|V_T|$.

A first-order quantitative expression for the ΔV_T associated with short-channel effects can be established using straightforward geometric arguments. Although highly simplified, the derivation to be presented is very informative in that it illustrates the general method of analysis. The derivation also indicates how parameters such as the source/drain junction depth enter into the specification of short-channel effects.

As previously established, for an ideal device

$$V_G = \phi_S + \frac{K_S}{K_O}x_o\mathscr{E}_S \tag{5.2}$$

Let Q_S be the total charge/cm^2 inside the semiconductor. Applying Gauss' law, it is readily established that

$$Q_S = -K_S\varepsilon_0\mathscr{E}_S \tag{5.3}$$

Combining Eqs. (5.2) and (5.3), we can therefore write

$$V_G = \phi_S - \frac{x_o Q_S}{K_O\varepsilon_0} = \phi_S - \frac{Q_S}{C_o} \tag{5.4}$$

when $V_G = V_T$, $\phi_S = 2\phi_F$, and $Q_S = Q_B$, where Q_B is the bulk or depletion-region charge per unit area of the gate. Specialized to the threshold point, Eq. (5.4) thus becomes

$$V_T = 2\phi_F - \frac{Q_B}{C_o} \qquad (5.5)$$

Next introducing

$$\Delta V_T \equiv V_T(\text{short channel}) - V_T(\text{long channel}) \qquad (5.6)$$

taking Q_{BL} and Q_{BS} to be the long-channel and short-channel depletion-region charges/cm^2, respectively, and making use of Eq. (5.5), we obtain

$$\Delta V_T = -\frac{1}{C_o}(Q_{BS} - Q_{BL}) = \frac{Q_{BL}}{C_o}\left(1 - \frac{Q_{BS}}{Q_{BL}}\right) \qquad (5.7)$$

To complete the derivation it is necessary to develop expressions for Q_{BL} and Q_{BS} in terms of the device parameters. Working toward this end, consider the short n-channel MOSFET pictured in Fig. 5.4. To simplify the analysis V_D is taken to be small or zero so that $W \cong W_T$ at all points beneath the central portion of the gate. The shaded areas in the figure identify those portions of the subgate region that are assumed to be controlled by the source and drain pn junctions. In a long-channel device, charge in the entire rectangular region of side-length L is balanced by charges on the gate and

$$Q_{BL} = -\frac{qN_A(ZLW_T)}{ZL} = -qN_A W_T \qquad (5.8)$$

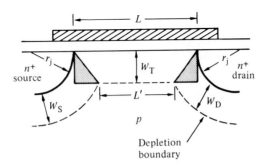

Fig. 5.4 Cross section of a MOSFET identifying parameters that enter into the short-channel analysis. The shaded portions of the subgate area are assumed to be controlled by the source and drain pn junctions. It is also assumed that $V_D \cong 0$.

ZLW_T is of course the depletion-region volume and ZL the area of the gate. In a short-channel device, the depletion-region charge controlled by the gate is confined to the trapezoidal region of side lengths L and L'. Thus

$$Q_{BS} = -\frac{qN_A\left[\frac{1}{2}(L + L')ZW_T\right]}{ZL} = -qN_A W_T\frac{L + L'}{2L} \tag{5.9}$$

Substituting the Q_{BL} and Q_{BS} expressions into Eq. (5.7) then gives

$$\Delta V_T = -\frac{qN_A W_T}{C_o}\left(1 - \frac{L + L'}{2L}\right) \tag{5.10}$$

Next, looking at the source region in Fig. 5.4 and assuming $W_S \cong W_T$, one deduces from geometrical considerations

$$(r_j + W_T)^2 = \left(r_j + \frac{L - L'}{2}\right)^2 + W_T^2 \tag{5.11}$$

which can be solved to obtain

$$L' = L - 2r_j\left[\sqrt{1 + \frac{2W_T}{r_j}} - 1\right] \tag{5.12}$$

Finally, eliminating L' in Eq. (5.10) using Eq. (5.12), we conclude

$$\boxed{\Delta V_T(\text{short channel}) = -\frac{qN_A W_T}{C_o}\frac{r_j}{L}\left(\sqrt{1 + \frac{2W_T}{r_j}} - 1\right)} \tag{5.13}$$

Although a first-order result, the ΔV_T given by Eq. (5.13) does exhibit the same parametric dependencies initially noted in the L_{min} discussion. Examing $\Delta V_T/V_T$(long-channel), which is the relevant quantity in gauging the importance of short-channel effects, one again finds the effects are minimized by reducing x_o, reducing r_j, and increasing N_A.

Narrow Width

The threshold voltage is also affected when the lateral width Z of a MOSFET becomes comparable to the channel depletion-width W_T. In enhancement-mode narrow-width devices, $|V_T|$ is found to monotonically *increase* with decreasing channel width Z. Note that the Z-dependence of the threshold voltage shift is opposite to the L-dependence. The narrow-width effect, however, is explained in much the same manner as the short-channel effect. Referring to the side view of a MOSFET in Fig. 5.5, note that the gate-controlled depletion region extends to the side, lying in part outside the Z-width of the gate. In wide-width devices, the gate controlled charge in the lateral region is totally negligible. In narrow-width devices, on the other hand, the lateral charge becomes comparable to the charge directly beneath the Z-width of the gate; that is, there is an increase in the effective charge/cm^2 being balanced by the gate charge. Thus, added gate charge is required to reach the start of inversion and $|V_T|$ increases.

Paralleling the short-channel derivation, a quantitative expression for ΔV_T associated with the narrow-width effect is readily established. If the lateral regions are assumed to be quarter-cylinders of radius W_T, the lateral volume controlled by the gate is $(\pi/2)W_T^2L$ and

$$Q_B(\text{narrow width}) = -\frac{qN_A\left(ZLW_T + \frac{\pi}{2}W_T^2L\right)}{ZL} = -qN_A W_T\left(1 + \frac{\pi}{2}\frac{W_T}{Z}\right) \quad (5.14)$$

Replacing Q_{BS} in Eq. (5.7) with the narrow-width Q_B, one rapidly concludes

$$\boxed{\Delta V_T(\text{narrow width}) = \frac{qN_A W_T}{C_o}\frac{\pi W_T}{2Z}} \quad (5.15)$$

This result confirms our initial assertion that the narrow-width effect becomes important when Z is comparable to W_T.

As a concluding point it should be noted that a combined-effect ΔV_T must be established for MOSFETs that have both a short channel and a narrow width. The short-

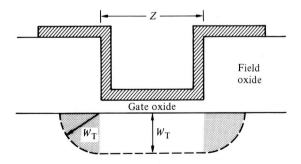

Fig. 5.5   Side view of a MOSFET used to explain and analyze the narrow-width effect.

channel and narrow-width ΔV_T's are not simply additive. The combined-effect ΔV_T and more exacting ΔV_T expressions for the individual effects can be found in the device literature.*

5.1.3 Parasitic BJT Action

Containing an oppositely doped region between the source and drain, the MOSFET bears a striking physical resemblance to a lateral bipolar junction transistor (BJT). Thus, with the distance between the source and drain in a modern MOSFET reduced to a value comparable to the base width in a bipolar transistor, it is not surprising that phenomena have been observed which are normally associated with the operation of BJTs.

One such phenomenon is source to drain *punch-through*. When the source and drain are separated by a few microns or less it becomes possible for the *pn* junction depletion regions around the source and drain to touch or punch-through as pictured in Fig. 5.6. When punch-through occurs, a significant change takes place in the operation of the MOSFET. Notably, the gate loses control of the subgate region except for a small portion of the region immediately adjacent to the $Si-SiO_2$ interface. The source-to-drain current is then no longer constrained to the surface channel, but begins to flow beneath the surface through the touching depletion regions. Analogous to the punch-through current in BJTs, this subsurface "space-charge" current varies as the square of the voltage applied between the source and drain. The $V_G > 0$ characteristics presented in Fig. 5.2 are examples of the $I_D \propto V_D^2$ current that results from source to drain punch-through.

As a practical matter, punch-through in small-dimension MOSFETs is routinely suppressed by increasing the doping of the subgate region and thereby decreasing the source/drain depletion widths. This can be accomplished by increasing the substrate doping. However, increasing the substrate doping has the adverse affect of increasing

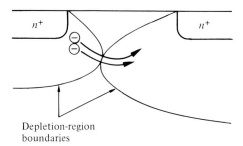

Depletion-region
boundaries

Fig. 5.6 Punch-through and space-charge current in a short-channel MOSFET.

---

*See, for example, T. A. DeMassa and H. S. Chien, "Threshold Voltage of Small-Geometry Si MOSFETs," *Solid-State Electronics*, **29**, 409 (1986).

parasitic capacitances. Consequently, it is common practice to perform a deep-ion implantation to selectively increase the doping of the subgate region.

Parasitic BJT action involving *carrier multiplication and regenerative feedback* leads to a second potentially significant perturbation of the MOSFET characteristics. There is a certain amount of carrier multiplication in the high-field depletion region near the drain in all MOSFETs. In long-channel devices the multiplication is negligible. In short-channel devices, however, carrier multiplication coupled with regenerative feedback can dramatically increase the drain current and place a reduced limit on the maximum operating V_D. Catastrophic failure may even occur under extreme conditions.

The multiplication and feedback mechanism operating in small-dimension MOSFETs is very similar to that which lowers BV_{ECO} in a BJT relative to BV_{BCO} (see Subsection 3.3.2 in Volume III). The basic mechanism is best described and explained with the aid of Fig. 5.7. The process is initiated by channel current entering the high-field region in the vicinity of the drain. Upon acceleration in the high-field region, a small percentage of the channel carriers gain a sufficient amount of energy to produce electron–hole pairs through impact ionization. For an *n*-channel device the added electrons drift into the drain, while the added holes are swept into the quasi-neutral bulk. Because the semiconductor bulk has a finite resistance, the current flow associated with the impact-generated holes gives rise to a potential drop between the depletion region edge and the back contact. This potential drop is of such a polarity as to forward bias the source *pn* junction. Forward biasing of the source *pn* junction in turn leads to electron injection from the source *pn* junction into the quasi-neutral bulk, an additional electron flow into the drain, increased carrier multiplication, and so forth. The process is stable as

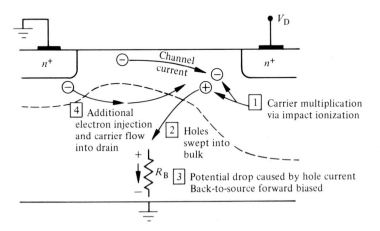

Fig. 5.7 Visualization of carrier multiplication and regenerative feedback, which can give rise to current enhancement in short-channel MOSFETs.

as long as the fractional increase in the drain current or the multiplication factor is less than $1/\alpha$, where α is the common-base gain of the parasitic BJT. At high enhanced currents there is the potential for excessive current flow through the device and device failure.

5.1.4 Hot-Carrier Effects

Oxide Charging

Oxide charging, or charge injection and trapping in the oxide, is a phenomenon that occurs in all MOSFETs. In the vicinity of the drain under operational conditions, channel carriers, and carriers entering the drain depletion region from the substrate, periodically gain a sufficient amount of energy to surmount the Si–SiO_2 surface barrier and enter the oxide. Neutral centers in the oxide trap a portion of the injected charge and thereby cause a charge build-up within the oxide. In long-channel devices, oxide charging is the well-known cause of "walk-out" — an increase in the drain breakdown over time in MOSFETs operated at large V_D biases. Unfortunately, the effects of oxide charging in short-channel MOSFETs are decidedly more serious. This is true because a larger percentage of the gated region is affected in the smaller devices. Specifically, significant changes in V_T and g_m can result from the oxide-charging phenomenon. Moreover, because oxide charging is cumulative over time, the phenomenon tends to limit the useful "life" of a device and must be minimized. A popular approach for minimizing hot-carrier effects, the formation of a lightly doped drain (LDD), is described in Subsection 5.2.1.

Velocity Saturation

In the conventional analysis of the long-channel MOSFET there is no theoretical limitation on the velocity that the carriers can attain in the surface channel. It is implicitly assumed the carrier velocities increase as needed to support the computed current. In reality, the carrier drift velocities inside silicon at $T = 300$ K approach a maximum value of $v_{dsat} \cong 10^7$ cm/sec when the accelerating electric field exceeds $\sim 3 \times 10^4$ V/cm for electrons and $\sim 10^5$ V/cm for holes (see Fig. 3.4 in Volume I). If, for example, $V_D = 5$ V and $L = 1$ μm, there will obviously be points in the MOSFET surface channel where the accelerating electric field is greater than or equal to 5×10^4 V/cm. Limitation of the channel current due to velocity saturation is clearly a possibility in short-channel devices.

Velocity saturation has two main effects on the observed characteristics. First, I_{Dsat} is significantly reduced. The modified I_{Dsat} is approximately described by

$$I_{Dsat} \cong ZC_o(V_G - V_T)v_{dsat} \tag{5.16}$$

Second, as can be inferred from Eq. (5.16) and as illustrated in Fig. 5.8, the saturation current exhibits an almost linear dependence on V_G–V_T as opposed to the conventional square-law dependence.

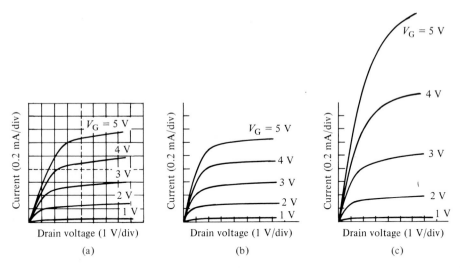

Fig. 5.8 Illustration of the effects of velocity saturation on the MOSFET I_D–V_D characteristics. (a) Experimental characteristics derived from a short-channel MOSFET with $L = 2.7$ μm, $x_o = 0.05$ μm, $r_j = 0.4$ μm, and N_A(substrate) $\cong 10^{15}/\text{cm}^3$. Comparative theoretical characteristics computed (b) including velocity saturation and (c) ignoring velocity saturation. [From K. Yamaguchi, *IEEE Trans. on Electron Devices*, **ED-26**, 1068, © 1979 IEEE].

Ballistic Transport

In the Chapter 3 description of carrier drift in the MOSFET surface channel, we implicitly assumed that carriers experienced numerous scattering events in traveling between the source and the drain. This is equivalent to assuming the channel length (L) is much greater than the average distance (l) between scattering events. Clearly, if the MOSFET channel length is reduced to a value comparable to l, fundamental revisions will be required in the analytical formalism. More importantly, however, if even smaller dimension structures could be built with $L < l$, a large percentage of the carriers would then travel from the source to the drain without experiencing a single scattering event. The envisioned projectilelike motion of the carriers is referred to as *ballistic transport*.

Observable ballistic effects are theoretically possible in GaAs structures where $L \leq 0.3$ μm. Somewhat shorter lengths are required in Si devices. Experimentally, both Si and GaAs FETs with channel lengths ~0.1 μm have been fabricated in research laboratories. Thus, FET structures with $L \leq l$ are attainable with modern-day technology.

Practically speaking, ballistic transport is of interest because it could lead to superfast devices. Not limited by scattering, the average velocity of carriers traversing the channel is expected to be greater than v_{dsat}. Unfortunately, there are other considerations, such as the operation of the injection source, which often limit device performance. Nonetheless, ballistic effects have been observed and ballistic transport will continue to be investigated.

5.2 SELECT STRUCTURE SURVEY

In discussing the MOSFET we made use of the basic enhancement-mode structure. Use of the basic structure allows one to focus on the development of concepts and the understanding of phenomena. Naturally, given the maturity of MOSFET technology, there exists a significant number of distinct device variations. Modifications to the basic structure have been implemented to solve specific problems or to enhance a specific device characteristic. It is also true that FET devices of all types fabricated in GaAs invariably take a different form than those fabricated in Si. A brief survey of select Si and GaAs FET structures has been included in this section to provide some feel for the variety that exists and the nature of the modifications. It should be emphasized that the surveyed device structures constitute only a sampling, with a bias toward structures likely to be encountered in the recent FET literature.

5.2.1 Si FETs

LDD Transistors*

As described in the preceding section, reduced dimension devices are more susceptible to hot-carrier effects. The field-aided injection and subsequent trapping of carriers in the gate oxide near the drain can lead to serious device degradation. The degrading effects are further worsened by the common practice of using bias voltages that have not been scaled down in proportion to the device dimensions. The lightly doped drain (LDD) structure shown in Fig. 5.9 helps to minimize hot-carrier effects. The feature of note is the lightly doped drain region between the end of the channel and the drain proper (the n^- region in Fig. 5.9). The reduced doping gradient in going from the channel to the drain proper lowers the $\mathscr{E}$-field in the vicinity of the drain and shifts the position of the peak $\mathscr{E}$-field toward the end of the channel. Carrier injection into the oxide is thereby reduced and oxide charging correspondingly minimized.

DMOS

A double-diffused MOSFET (DMOS) structure is pictured in Fig. 5.10. The structure is distinctive in that the channel region is formed, and the channel length specified, by the difference in the lateral extent of two impurity profiles. A p-type dopant (e.g., boron) and an n-type dopant (e.g., phosphorus) are admitted and diffused into the Si

---

*For a comprehensive review of LDD and similar structures see J. J. Sanchez, K. K. Hsueh, and T. A. DeMassa, "Drain-Engineered Hot-Electron Resistant Device Structures: A Review," *IEEE Trans. on Electron Devices,* **36**, 1125 (June 1989).

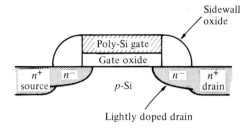

Fig. 5.9 Cross section of a lightly doped drain (LDD) structure.

through the same oxide mask opening. The p-type dopant, which is introduced first, diffuses slightly deeper and farther to the side than the n-type dopant. The result is the simultaneous formation of the source and channel regions as clearly shown in the mag-

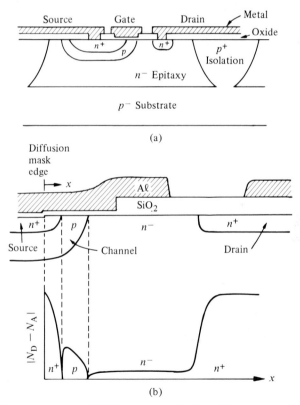

Fig. 5.10 (a) Cross section of a DMOS structure. (b) Magnified cross section of the channel region and lateral-doping profile. [From M. D. Pocha, A. G. Gonzalez, and R. W. Dutton, *IEEE Trans. on Electron Devices,* **ED-21**, 778, © 1974 IEEE.]

nified cross section of Fig. 5.10(b). The most important physical characteristic of DMOS is a short-channel length (~1 μm) that can be established without using small-dimension lithographic masks. The DMOS structure thus boasts high-frequency operation, which is combined with a high drain-breakdown voltage. It has been used in high-frequency analog applications and high-voltage/high-power circuits. Although first introduced in the early 1970s, variations of the DMOS structure, notably power-DMOS structures, continue to be developed.

Buried-Channel MOSFET

Figure 5.11 shows the cross section of a buried-channel MOSFET and the approximate subgate-doping profile inside the transistor. The unique structural feature, a surface layer beneath the gate with the same doping as the source and drain islands, is typically formed by ion implantation. With a *pn* junction bottom gate and an MOS top gate, the buried-channel MOSFET is physically and functionally a hybrid JFET/MOSFET structure. It can be designed to function as a depletion-mode or enhancement-mode device depending on the thickness and doping of the surface layer. The buried-channel MOSFET gets its name from the fact that channel conduction can be made to take place away from the oxide-semiconductor interface. This leads to inherently higher carrier mobilities, a reduced interfacial trap interaction, and a decreased sensitivity to hot-carrier effects.

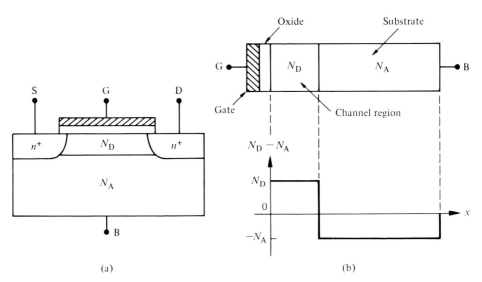

(a)   (b)

Fig. 5.11 (a) Cross section of a buried-channel MOSFET. (b) Approximate subgate doping profile. [Adapted from M. J. Van der Tol and S. G. Chamberlain, *IEEE Trans. on Electron Devices,* **36**, 670, © 1989 IEEE.]

SOI Structures

The term "silicon-on-insulator" (SOI) is used to describe structures where devices are fabricated in Si-layers formed *over* an insulating film or substrate. Traditional advantages, which have sustained interest in SOI structures, include low parasitic capacitances and potentially ultra-high speed related to the dielectric isolation of devices, improved radiation hardness, and the exciting possibility of constructing stacked or "3D" device configurations. The first realization of SOI structures involved crystalline films of Si deposited on properly oriented sapphire substrates (SOS). Subsequently, laser annealing techniques were used to crystallize amorphous Si films deposited on insulators such as SiO_2 and Si_3N_4. In both of these approaches the Si film quality has prevented the realization of the promised potential. However, a recently introduced SOI method, epitaxial layer overgrowth (ELO), appears to have solved the film quality problem. A hole is opened in the insulator to expose a small section of the underlying crystalline-Si substrate. With the exposed substrate acting as a seed, epitaxial Si is then grown up through the hole and sideways over the insulator. The adjacent cross sections of a prototype 3D CMOS inverter incorporating a *p*-channel transistor formed in an ELO film and an individual SOI-PMOS transistor are reproduced in Fig. 5.12.

It should be noted that SOI-MOSFETs fabricated in ultra-thin (~100 Å) Si-films are expected to provide additional performance advantages. The recently recognized advantages include relative insensitivity to small-dimension and hot-carrier effects, a steeper slope to the subthreshold transfer characteristics, and immunity to the kink effect (an undesirable "kink" in the I_D–V_D characteristics that plagues devices fabricated in thicker SOI films).

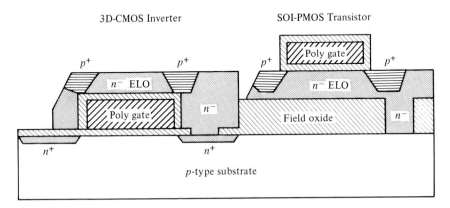

Fig. 5.12 Cross sections of sample silicon-on-insulator structures containing ELO films. [From J. A. Friedrich, M. Kastelic, G. W. Neudeck, and C. G. Takoudis, *J. Appl. Phys.*, **65**, 1713 (February 1989).]

5.2.2   GaAs FETs

D-MESFET and E-MESFET

As noted in the General Introduction, the electron transport properties of GaAs are superior to those of silicon. GaAs n-channel FETs are therefore expected to achieve higher operating speeds. Other GaAs device advantages that might be exploited are lower noise levels, a higher tolerance to radiation, and a potentially higher temperature of operation. Unfortunately, GaAs technology lacks an insulator comparable to thermally grown SiO_2; presently there is no true GaAs equivalent of the silicon-based MOSFET. As an alternative, metal-semiconductor FETs have been developed for use in GaAs ICs. The center of most GaAs commercial and defense program activity, the depletion-mode or D-MESFET pictured in Fig. 5.13(a) has achieved production-line status. Consistent with the depletion-mode designation, the D-MESFET is a normally-on n-channel device with a negative threshold voltage. $V_T = -0.5$ V to -1.0 V and channel lengths on the order of 1 μm are fairly typical. The enhancement mode or E-MESFET [Fig. 5.13(b)], on the other hand, is fabricated so that the built-in voltage associated with the metal-semiconductor contact is sufficient to totally deplete the implanted channel. It is therefore necessary to forward bias the metal-semiconductor gate in order to turn the E-MESFET on. Because of the forward bias operation of the gate, the voltage swing is extremely limited (<0.5 V) and proper IC operation can tolerate

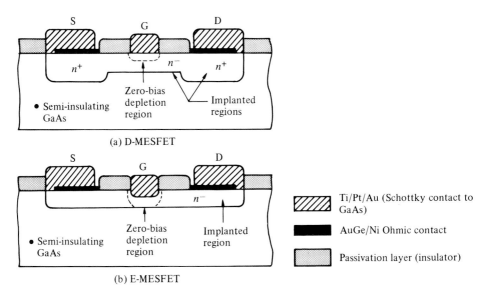

Fig. 5.13 GaAs MESFET structures. (a) Depletion-mode or D-MESFET; (b) Enhancement-mode or E-MESFET.

only small (~25 mV) device-to-device variations in the threshold voltage. Thus, although the E-MESFET offers a number of circuit-related advantages, fabrication difficulties have limited utilization of the structure.

MODFET (HEMT)

The last structure to be examined is the modulation doped field effect transistor (MODFET). Another widely used name for the same structure is the high electron mobility transistor (HEMT). A perspective view of a MODFET or HEMT intended for use in ICs was presented in Fig. 7 of the General Introduction; a simplified cross section of the basic device structure is shown in Fig. 5.14(a). At the present time, the GaAs and AlGaAs layers in the structure are typically formed using molecular beam epitaxy (MBE).* The alloy semiconductor AlGaAs has essentially the same lattice constant as GaAs and can therefore be grown as a near-perfect extension of the GaAs crystalline lattice.

Physically and functionally the MODFET is a MOSFET-like device, with the AlGaAs assuming the role of quasi-insulator. As pictured in Fig. 5.14(b), AlGaAs has a wider band gap than GaAs, which gives rise to an electron containment barrier at the AlGaAs–GaAs interface. The AlGaAs is specifically doped to create an inversion or accumulation layer of electrons at the GaAs surface. (The conducting electron layer is usually referred to as a two-dimensional gas in MODFET terminology.) The AlGaAs layer is also made sufficiently thin so that it is totally depleted by the Schottky-gate barrier potential under equilibrium conditions. The resultant structure is *electrostatically* similar to a MOSFET with sodium ions distributed throughout the oxide. It is important to note that, unlike the MESFET structures, the active GaAs layer in the MODFET is nominally undoped. Thus there is only minimal scattering from residual (unintentional) dopant impurities in the GaAs surface channel. Very high electron mobilities are observed at room temperature, and an even greater mobility enhancement relative to MESFETs is obtained at liquid nitrogen temperatures.

The MODFET is considered to be a strong contender for use in high-speed logic circuits of the 1990s. There are presently threshold-voltage problems similar to that of the E-MESFET, and the MBE process used to grow the GaAs and AlGaAs layers is not viewed as a production-line technique. Device development of course continues and, at the very least, the MODFET appears to be a forerunner of things to come.

PROBLEMS

5.1 Answer the following questions as concisely as possible.

(a) Name the three most commonly cited indications of short-channel effects.

(b) Relative to threshold-voltage modification, how do the short-channel and narrow-width effects differ? In what ways are they alike?

---

*See Subsection 6.4.4 in Volume V for a short description and discussion of MBE.

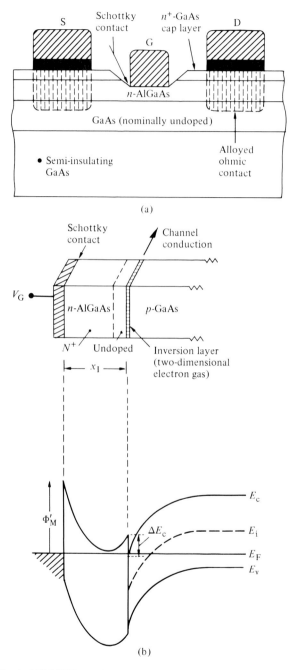

Fig. 5.14   The basic MODFET or HEMT structure. (a) Simplified cross section. (b) Section through the transistor beneath the gate and the associated energy band diagram. [Part (b) from R. F. Pierret and M. S. Lundstrom, *IEEE Trans. on Electron Devices,* **ED-31,** 383, © 1984 IEEE.]

(c) Name two BJT-like phenomena that have been observed in short-channel MOSFETs.

(d) Why is the oxide charging associated with hot-carrier effects more important in short-channel devices?

(e) Indicate what is meant by "ballistic transport."

(f) What is the major (most significant) physical difference between a D-MESFET and an E-MESFET?

(g) What do ELO, HEMT, and SOI stand for?

5.2 Briefly indicate the unique physical feature of the following transistor structures:

(a) LDD transistors

(b) DMOS

(c) Buried-channel MOSFET

(d) SOI structures.

5.3 (a) Making use of the parameters supplied in the figure caption, compute the expected L_{min} for the Fig. 5.3 device when $V_D = 0.125$ V. Assume n^+-p drain and source step junctions with a built-in voltage of $V_{bi} = 0.92$ V and that the source and back are grounded. Comment on your computational result.

(b) Utilizing Eq. (5.13), what is the ΔV_T expected for the Fig. 5.3 device when $L = 1$ μm and $V_D = 0.125$ V?

(c) Can Eq. (5.13) be applied to compute an expected ΔV_T for comparison with the $V_D = 4$ V data? Explain.

5.4 Read the following journal articles and prepare a one to two page summary of each article.

(a) J. J. Sanchez, K. K. Hseuh, and T. A. DeMassa, "Drain-Engineered Hot-Electron Resistant Device Structures: A Review," *IEEE Trans. on Electron Devices,* **36,** 1125 (June 1989).

(b) M. J. Van der Tol and S. G. Chamberlain, "Potential and Electron Distribution Model for the Buried-Channel MOSFET," *IEEE Trans. on Electron Devices,* **36,** 670 (April 1989).

(c) T. A. DeMassa and H. S. Chien, "Threshold Voltage of Small-Geometry Si MOSFETs," *Solid-State Electronics,* **29,** 409 (1986).

(d) C. G. Kirkpatrick, "Making GaAs Integrated Circuits," *Proc. IEEE,* **76,** 792 (July 1988).

Suggested Readings

General Comment: Listed below are six texts that, in the author's opinion, nicely supplement or enhance the material presented in this volume. The texts contain additional references that might also be consulted.

[1] R. C. Jaeger, *Introduction to Microelectronic Fabrication,* Volume V in the Modular Series on Solid State Devices, G. W. Neudeck and R. F. Pierret editors. Reading Mass.: Addison-Wesley, 1988. This volume provides useful information about fabrication details and device design. Chapter 9 is specifically devoted to MOS process integration.

[2] D. G. Ong, *Modern MOS Technology, Processes, Devices, & Design.* New York: McGraw-Hill, 1984. An easy to read book that is applications oriented and totally devoted to MOS devices.

[3] D. K. Schroder, *Advanced MOS Devices,* Volume VII in the Modular Series on Solid State Devices, R. F. Pierret and G. W. Neudeck editors. Reading Mass.: Addison-Wesley, 1987. A higher-level extension of Volume IV, the Schroder volume is recommended for the reader interested in learning about MOS memory cells, CCDs and/or more about MOS devices in general.

[4] B. G. Streetman, *Solid State Electronic Devices,* 3rd edition. Englewood Cliffs, N.J.: Prentice-Hall, 1990. See Chapter 8. This is a good general-purpose text, providing a different viewpoint on some of the topics covered in Volume IV.

[5] S. M. Sze, *Semiconductor Devices, Physics and Technology.* New York: Wiley, 1985. See Chapters 5 and 12. Supplemented by excellent figures and graphs, the Sze text provides a concise coverage of the subject matter with special attention to fabrication details.

[6] E. S. Yang, *Microelectric Devices.* New York: McGraw-Hill, 1988. See Chapters 8–13. Yang provides an alternative coverage of the subject matter which is quite readable and informative.

Appendix A
Exercises

EXERCISE 1.1

Q: Show that the source and drain resistances at the ends of the active channel (R_S and R_D in Fig. 1.12) are appropriately taken into account by replacing V_D with $V_D - I_D(R_S + R_D)$ and V_G with $V_G - I_D R_S$ in Eqs. (1.9), (1.12), and (1.13).

A: With the voltage drops at the channel ends taken into account, the channel voltages at $y = 0$ and $y = L$ are $V(0) = I_D R_S$ and $V(L) = V_D - I_D R_D$, repectively. Inserting the revised voltage limits into Eq. (1.4), and likewise modifying Eq. (1.5), we find

$$I_D = \frac{2qZ\mu_n N_D a}{L} \int_{I_D R_S}^{V_D - I_D R_D} \left[1 - \frac{W(V)}{a} \right] dV \tag{1.5'}$$

Since the Eq. (1.8) expression for $W(V)/a$ remains unchanged, the integral in Eq. (1.5') is readily evaluated. The integration yields

$$I_D = \frac{2qZ\mu_n N_D a}{L} \left\{ V_D - I_D(R_S + R_D) - \frac{2}{3}(V_{bi} - V_P) \left[\left(\frac{V_D - I_D R_D + V_{bi} - V_G}{V_{bi} - V_P} \right)^{3/2} \right. \right.$$
$$\left. \left. - \left(\frac{I_D R_S + V_{bi} - V_G}{V_{bi} - V_P} \right)^{3/2} \right] \right\} \tag{1.9'}$$

Turning next to the modification of Eq. (1.12), we note that when $V_D = V_{Dsat}$, $W \to a$, $V(L) = V_{Dsat} - I_{Dsat} R_D$, and from Eq. (1.6),

$$a = \left[\frac{2K_S \varepsilon_0}{qN_D} (V_{bi} + V_{Dsat} - I_{Dsat} R_D - V_G) \right]^{1/2} \tag{1.11'}$$

However,

$$a = \left[\frac{2K_S \varepsilon_0}{qN_D} (V_{bi} - V_P) \right]^{1/2} \tag{1.7}$$

and clearly,

$$V_{Dsat} - I_{Dsat} R_D = V_G - V_P \tag{1.12'}$$

Finally, setting $V_D = V_{Dsat}$ and $I_D = I_{Dsat}$ in Eq. (1.9'), and simplifying the result using Eq. (1.12'), one obtains

$$I_{Dsat} = \frac{2qZ\mu_n N_D a}{L} \left\{ V_G - V_P - I_{Dsat} R_S - \frac{2}{3}(V_{bi} - V_P) \times \left[1 - \left(\frac{I_{Dsat} R_S + V_{bi} - V_G}{V_{bi} - V_P} \right)^{3/2} \right] \right\} \tag{1.13'}$$

Note that replacing V_D by $V_D - I_D(R_S + R_D)$ and V_G by $V_G - I_D R_S$ in Eqs. (1.9), (1.12), and (1.13) does indeed yield Eqs. (1.9'), (1.12'), and (1.13'), respectively.

Before concluding, a few comments are in order concerning these results. First, it is obvious that the primed equations cannot be solved for I_D or I_{Dsat} as a closed-form function of V_D and V_G. Nevertheless, numerical iteration techniques can be used to readily calculate the revised current-voltage characteristics. Second, the answer to this question could have been established by simply changing the integration variable in Eq. (1.5') to $V' = V - I_D R_S$, and subsequently requiring the original and the revised versions of Eq. (1.5) to have the same general form. (The change-of-variables approach, however, is less informative.)

EXERCISE 1.2

Q: Using the equivalent circuit of Fig. 1.14, and assuming the J-FET is saturation biased, show that the source and drain resistances at the ends of the active channel give rise to a reduced effective transconductance

$$g'_m = \frac{g_m}{1 + g_m R_S}$$

A: Under saturation conditions $g_d \to 0$. Open-circuiting g_d and adding R_S and R_D to the proper nodes in Fig. 1.14 yields the working equivalent circuit displayed in Fig. E1.2. As deduced from Fig. E1.2,

$$i_d = g_m v'_g = g_m(v_g - i_d R_S)$$

Thus

$$i_d = \frac{g_m}{1 + g_m R_S} v_g \equiv g'_m v_g$$

where

$$g'_m = \frac{g_m}{1 + g_m R_S}$$

It is interesting to note that R_D does not affect the transconductance if $g_d = 0$. R_S, on the other hand, causes a decrease in the effective transconductance of the J-FET if $g_m R_S$ is comparable to unity.

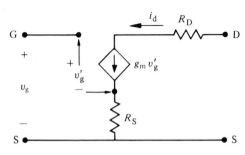

Fig. E1.2

EXERCISE 2.1

Q: For the ϕ_F, ϕ_S parameter sets listed below, first indicate the specified biasing condition and then draw the energy band diagram and block charge diagram that characterize the static state of the system. Assume the MOS structure to be ideal.

(a) $\dfrac{\phi_F}{kT/q} = 12$, $\dfrac{\phi_S}{kT/q} = 12$

(b) $\dfrac{\phi_F}{kT/q} = -9$, $\dfrac{\phi_S}{kT/q} = 3$

(c) $\dfrac{\phi_F}{kT/q} = -9$, $\dfrac{\phi_S}{kT/q} = -18$

(d) $\dfrac{\phi_F}{kT/q} = 15$, $\dfrac{\phi_S}{kT/q} = 36$

(e) $\dfrac{\phi_F}{kT/q} = -15$, $\dfrac{\phi_S}{kT/q} = 0$

A:

Set	Doping	Biasing condition	Energy band diagram	Block charge diagram
(a)	p	Depletion		
(b)	n	Accumulation		
(c)	n	Depl/Inv Transition		
(d)	p	Inversion		
(e)	n	Flat Band		

EXERCISE 2.2

Q: An MOS-C is maintained at $T = 300$ K, $x_o = 0.1$ μm, and the Si doping is $N_A = 10^{15}/cm^3$. Compute:

(a) ϕ_F (in kT/q units and in volts)

(b) W when $\phi_S = \phi_F$

(c) $\mathscr{E}_S$ when $\phi_S = \phi_F$

(d) V_G when $\phi_S = \phi_F$

A: (a)

$$\frac{\phi_F}{kT/q} = \ln(N_A/n_i) = \ln\left(\frac{10^{15}}{1.18 \times 10^{10}}\right) = 11.35$$

$$\phi_F = 11.35(kT/q) = (11.35)(0.0259) = 0.294 \text{ V}$$

(b) Using Eq. (2.15),

$$W = \left[\frac{2K_S\varepsilon_0}{qN_A}\phi_F\right]^{1/2} = \left[\frac{2(11.8)(8.85 \times 10^{-14})(0.294)}{(1.6 \times 10^{-19})(10^{15})}\right]^{1/2} = 0.620 \text{ } \mu m$$

(c) Evaluating Eq. (2.12) at $x = 0$ yields $\mathscr{E}_S$. Thus

$$\mathscr{E}_S = \frac{qN_A}{K_S\varepsilon_0}W = \frac{(1.6 \times 10^{-19})(10^{15})(6.20 \times 10^{-5})}{(11.8)(8.85 \times 10^{-14})} = 9.50 \times 10^3 \text{ V/cm}$$

(d) Substituting into Eq. (2.26) gives

$$V_G = \phi_F + \frac{K_S}{K_O} x_o \mathcal{E}_S \quad \ldots \mathcal{E}_S \text{ evaluated at } \phi_F$$

$$= 0.294 + \frac{(11.8)(10^{-5})(9.50 \times 10^3)}{3.9}$$

$$= 0.581 \text{ V}$$

Comment: The manipulations and results in this exercise are fairly representative. In kT/q units, $|\phi_F|$ typically ranges between 9 and 15 at $T = 300$ K. For the nondegenerate dopings used in MOS devices one also expects $|\phi_F|$ to be less than one-half a Si band gap expressed in volts (<0.56 V at $T = 300$ K). As required, the calculated W is less than W_T associated with the given doping (see Fig. 2.9). The only possible surprise is the size of the surface electric field — $\mathcal{E}_S \sim 10^4$ V/cm. Finally, the device parameters assumed in this exercise are identical to those used in constructing Fig. 2.10. The computed V_G of course agrees with the value read from the Fig. 2.10 plot.

EXERCISE 2.3

Q: Complete the following table making use of the ideal-structure $C-V$ characteristic and the block charge diagrams included in Figure E2.3. For each of the biasing conditions named in the table, employ letters (a–g) to identify the corresponding bias point or points on the ideal MOS-C $C-V$ characteristic. Likewise, using a number (1–5) identify the block charge diagram associated with each of the biasing conditions.

Bias condition	Capacitance (a–g)	Block charge diagram (1–5)
Accumulation		
Depletion		
Inversion		
Flat band		
Depl/inv transition		

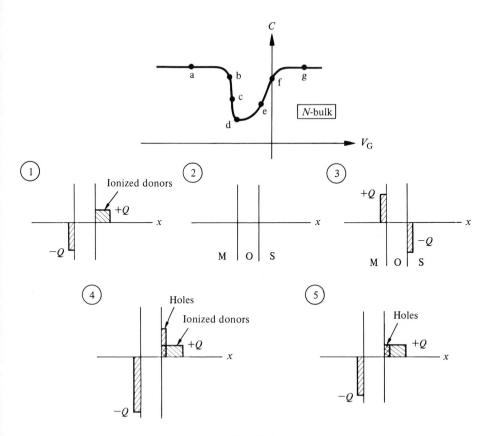

Fig. E2.3

EXERCISE 2.4

Q: The experimental C–V characteristic shown in Fig. E2.4 was observed under the following conditions: The dc bias was changed very slowly from point (1) to point (2). At point (2) the V_G sweep rate was increased substantially. Upon arriving at point (3) the sweep was stopped and the capacitance decayed to point (4). Qualitatively explain the observed characteristic.

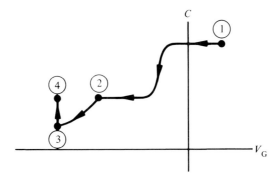

Fig. E2.4

A: In going from point (1) to point (2) the semiconductor clearly equilibrates at each dc bias point because of the very slow speed of the ramp. One therefore observes the standard high-frequency C–V characteristic. With the ramp rate increased at point (2), the semiconductor is no longer able to equilibrate and is consequently driven into deep depletion. W becomes larger than W_T and C decreases below the minimum high-frequency equilibrium value. When the sweep is stopped and the capacitance increases from point (3) to point (4), equilibrium is systematically restored inside the device through the generation of minority carrier holes in the near-surface region. As the holes are generated they add to the inversion layer, the depletion width correspondingly decreases, and C increases back to the equilibrium high-freqency value.

EXERCISE 3.1

Q: Suppose the gate of an ideal n-channel MOSFET is connected to the drain making $V_G = V_D$. Utilizing the square-law results, sketch I_D versus V_D ($V_D \geq 0$).

A: In an ideal n-channel MOSFET $V_T > 0$. Thus, with the gate and drain tied together,

$$V_D = V_G > V_G - V_T$$

but

$$V_{Dsat} = V_G - V_T$$

giving

$$V_D > V_{Dsat}$$

The device is *always* saturation biased. Noting $I_D = 0$ if $V_D = V_G < V_T$ and using Eq. (3.20), we conclude

$$I_D = \begin{cases} 0 & \dots V_D = V_G < V_T \\ \dfrac{Z\,\bar{\mu}_n C_o}{2L}(V_G - V_T)^2 = \dfrac{Z\bar{\mu}_n C_o}{2L}(V_D - V_T)^2 & \dots V_D > V_T \end{cases}$$

The required I_D–V_D sketch is shown in Fig. E3.1.

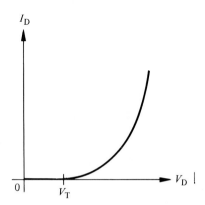

Fig. E3.1

EXERCISE 3.2

Q: A complementary pair of ideal *n*-channel and *p*-channel MOSFETs is to be designed so that the devices exhibit the same g_m and f_{max} when equivalently biased and operated at $T = 300$ K. The structural parameters of the *n*-channel device are $Z = 50$ μm, $L = 5$ μm, $x_o = 0.05$ μm, and $N_A = 10^{15}/\text{cm}^3$. The *p*-channel device has the same oxide thickness and doping concentration, but, because of the lower hole mobility, must have different gate dimensions. Determine the required Z and L of the *p*-channel device. Assume the effective mobility of carriers in both devices is one-half the bulk mobility.

A: $|V_T|$ is the same in ideal *p*-channel and *n*-channel MOSFETs with the same x_o and bulk doping concentration. Thus, with the devices also equivalently biased, one concludes from Table 3.1 that the same g_m's will result if

$$\frac{Z_p}{L_p}\bar{\mu}_p = \frac{Z_n}{L_n}\bar{\mu}_n$$

where the subscripts indicate the channel type. This same conclusion is reached whether one uses the square-law theory or bulk-charge theory and whether the devices are biased below or above pinch-off.

Next, examining the first form of Eq. (3.33), we again quite generally conclude that $C_O(p$-channel) must equal $C_O(n$-channel) for the f_{max} values to be the same. Since $C_O = K_O \varepsilon_0 ZL/x_o$, we therefore require

$$Z_p L_p = Z_n L_n$$

Substituting Z_p from the first relationship into the second relationship and simplifying, one obtains

$$L_p = \sqrt{\frac{\overline{\mu_p}}{\overline{\mu_n}}} \, L_n$$

and

$$Z_p = \frac{Z_n L_n}{L_p} = \sqrt{\frac{\overline{\mu_n}}{\overline{\mu_p}}} \, Z_n$$

The mobilities deduced from Fig. 3.5 in Volume I are $\overline{\mu}_n = \mu_n/2 = 673$ cm^2/V-sec and $\overline{\mu}_p = \mu_p/2 = 229$ cm^2/V-sec.* Thus the required p-channel device dimensions are $Z = \sqrt{673/229} \times 50 = 85.7$ μm and $L = \sqrt{229/673} \times 5 = 2.92$ μm.

---

*The carrier mobilities in Fig. 3.5, Volume I, are actually majority carrier mobilities — the electron mobility in N_D-doped Si and the hole mobility in N_A-doped Si. We are assuming here that the carrier mobilities are the same in equivalently doped Si of the opposite doping type.

EXERCISE 4.1

Q: It is possible to build an MOS-C with a $\phi_{MS} = 0$ through the proper choice of gate material and Si doping concentration. Restricting the Si doping to be in the range $10^{14}/\text{cm}^3 \leq N_A$ or $N_D \leq 10^{17}/\text{cm}^3$, and assuming operation at $T = 300$ K, identify the gate-material/doping-concentration combination(s) that gives rise to a $\phi_{MS} = 0$. Employ the $\Phi'_M - \chi'$ values given in Table 4.1.

A: Noting that

$$(E_c - E_F) = E_c - E_i + (E_i - E_F)_\infty$$

$$\cong \frac{E_G}{2} - kT \ln\left(\frac{N_D}{n_i}\right) \quad \ldots n\text{-type Si}$$

$$\cong \frac{E_G}{2} + kT \ln\left(\frac{N_A}{n_i}\right) \quad \ldots p\text{-type Si}$$

and employing $kT = 0.0259$ eV, $E_G = 1.12$ eV, and $n_i = 1.18 \times 10^{10}/\text{cm}^3$, one calculates

$$0.15 \text{ eV} \leq (E_c - E_F)_\infty \leq 0.33 \text{ eV} \qquad \ldots \text{if } 10^{14}/\text{cm}^3 \leq N_D \leq 10^{17}/\text{cm}^3$$

$$0.79 \text{ eV} \leq (E_c - E_F)_\infty \leq 0.97 \text{ eV} \qquad \ldots \text{if } 10^{14}/\text{cm}^3 \leq N_A \leq 10^{17}/\text{cm}^3$$

Since $\phi_{MS} = (1/q)[\Phi'_M - \chi' - (E_c - E_F)_\infty]$, to achieve a $\phi_{MS} = 0$ clearly requires

$$0.15 \text{ eV} \leq \Phi'_M - \chi' \leq 0.33 \text{ eV} \qquad \ldots \text{or} \ldots \qquad 0.79 \text{ eV} \leq \Phi'_M - \chi' \leq 0.97 \text{ eV}$$

Examining Table 4.1 we find that the only gate material that meets the general requirement is Au with a $\Phi'_M - \chi' = 0.82$ eV.

The specific doping of the Au p-Si MOS-C exhibiting a $\phi_{MS} = 0$ must be such that

$$(E_c - E_F)_\infty = \Phi'_M - \chi' = 0.82 \text{ eV}$$

or

$$(E_i - E_F)_\infty = 0.26 \text{ eV}$$

and

$$N_A = n_i e^{(E_i - E_F)_\infty/kT} = 2.70 \times 10^{14}/\text{cm}^3$$

Thus

Au gate; $N_A = 2.70 \times 10^{14}/\text{cm}^3 \Rightarrow$ MOS-C with $\phi_{MS} = 0$

EXERCISE 4.2

Q: Consider the MOS-C energy band diagram shown in Fig. E4.2. One concludes from the diagram that (choose one):

(a) $Q_M = 0$ or $Q_M \neq 0$? Explain.

(b) $Q_F = 0$ or $Q_F \neq 0$? Explain.

A: There is a very important point to this exercise; namely, the energy band diagram is modified when a significant number of charge centers are positioned in the oxide and/or at the oxide-semiconductor interface.

(a) $\boxed{Q_M \neq 0}$. If there is no charge in the oxide, if $\rho_{ox} = 0$, then $\mathscr{E}_{ox} = $ constant and the oxide energy bands are a linear function of position. However, if $\rho_{ox} \neq 0$, $\mathscr{E}_{ox}$ becomes a function

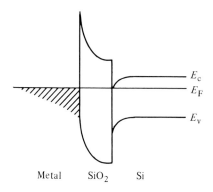

Fig. E4.2

of position and the oxide energy bands in turn exhibit curvature. A concave curvature as pictured in Fig. E4.2 is indicative of a significant positive charge, alkali ions, in the oxide.

(b) $\boxed{Q_F \neq 0}$. The normal component of the D-field, where $D = \varepsilon\mathscr{E}$, must be continuous if there is no plane of charge at an interface between two dissimilar materials (see Subsection 2.3.2). When a plane of charge does exist, there is a discontinuity in the D-field equal to the charge/cm^2 along the interface. Note from Fig. E4.2 that the slope of the bands is zero and therefore $\mathscr{E} = (1/q)(dE_c/dx) = 0$ on the oxide side of the interface. On the semiconductor side of the interface, $\mathscr{E}$ is decidedly nonzero and positive. Thus, there must be a plane of charge at the interface. For the pictured situation, we in fact require $Q_{interface} = K_S\varepsilon_0\mathscr{E}_S$ and the interface charge must be positive. In real devices alkali ions always give rise to a spread-out volume charge; that is, alkali ions are an unlikely source of the interface charge. The fixed charge, on the other hand, closely approximates a plane of positive charge at the Si–SiO$_2$ interface. We conclude $Q_F \neq 0$.

Although a conclusion has been reached, we need to address an apparent inconsistency. In this exercise we have indicated the fixed charge will cause a discontinuity in the interfacial D-field. However, in deriving Eq. (4.11), the D-field was *explicitly* assumed to be continuous across the Si–SiO$_2$ interface. Equation (4.11) in turn was used to establish the ΔV_G(fixed charge) expression. This apparent inconsistency is resolved if the mathematical development is examined carefully. To be precise, by including Q_F in ρ_{ox}, we are actually taking the fixed charge to be slightly inside the oxide. The discontinuity then occurs at $x = x_o^-$ instead of exactly at $x = x_o$. Whether the discontinuity occurs exactly at the interface or an imperceptible distance into the oxide cannot be detected physically, and clearly does not affect the mathematical results. Finally, it should be noted that, in general, a D-field discontinuity at the Si–SiO$_2$ interface can arise from other sources of interface charge including, for example, the interfacial trap charge discussed in Subsection 4.2.4.

EXERCISE 4.3

Q: Thermal oxidation of a (111)-oriented silicon substrate is followed by an N$_2$ anneal at 1000° C for a time sufficient to achieve a steady-state condition. After depositing Al the structure is next postmetallization annealed. The completed MOS-C is found to be stable under bias-temperature stressing. Determine the expected flat band voltage of the MOS-C if $T = 300$ K, $x_o = 250$ Å, and $N_A = 10^{15}/cm^3$.

A: (i) Figure 4.3 indicates $\phi_{MS} = -0.89$ V for an Al–SiO$_2$–Si structure with $N_A = 10^{15}/cm^3$ maintained at $T = 300$ K.

(ii) The nitrogen anneal minimizes the fixed charge making $Q_F/q = 2 \times 10^{11}/cm^2$ as deduced from Fig. 4.9(b). The fixed charge gives rise to a voltage shift

$$\Delta V_G \left(\begin{matrix} \text{fixed} \\ \text{charge} \end{matrix} \right) = -\frac{Q_F}{C_o} = -q \frac{x_o}{K_o \varepsilon_0} \frac{Q_F}{q} = -\frac{(1.6 \times 10^{-19})(2.5 \times 10^{-6})(2 \times 10^{11})}{(3.9)(8.85 \times 10^{-14})}$$

$$= -0.23 \text{ V}$$

(iii) We expect ΔV_G(interfacial traps) $\cong 0$ because of postmetallization annealing.

(iv) We know $Q_M \cong 0$ because the device is stable under bias-temperature stressing.

Thus

$$V_{FB} \cong \phi_{MS} - \frac{Q_F}{C_o} = -1.12 \text{ V}$$

EXERCISE 4.4

Q: The C_G–V_G ($V_D = 0$) characteristic derived from an n-channel MOSFET is pictured in Fig. E4.4.

(a) What is the threshold voltage, V_T, of the transistor? Explain how you arrived at your answer.

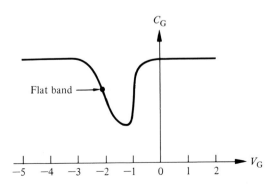

Fig. E4.4

(b) Is the MOSFET a depletion mode or an enhancement mode device? Explain.

(c) Sketch the general form of the I_D-V_D characteristics expected from the device, specifically labeling those characteristics corresponding to $V_G = -2, -1, 0, 1$, and 2 V.

(d) Given $\phi_{MS} = -1$ V, $Q_{IT} = 0$, and the fact that the device is stable under bias-temperature stressing, how do you explain the observed flat band voltage of -2.2 V?

(e) If the MOSFET is doped such that $\phi_F = 0.3$ V, what substrate bias (V_{BS}) must be applied to attain a $V_{GSlat\ threshold} = +1$ V?

A: (a) $\boxed{V_T \cong -1\ V}$. In progressing toward positive biases, the $C-V$ curve increases toward $C = C_O$ at roughly the inversion–depletion transition point. V_G at the inversion–depletion transition point is of course the threshold voltage in MOSFETs.

(b) $\boxed{\text{Depletion Mode}}$. With $V_T = -1$ V, the MOSFET is clearly "on" or conducting when $V_G = 0$.

(c) The required characteristics are sketched below.

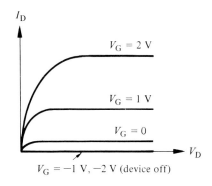

$V_G = -1$ V, -2 V (device off)

(d) Since the device is stable under bias-temperature stressing, one concludes $Q_M \cong 0$. With $Q_M = 0$, $Q_{IT} = 0$, and ϕ_{MS} unable to account for the entire flat band shift, the residual shifting can only be attributed to the $\boxed{\text{fixed charge}}$.

(e) As derived in the text

$$\Delta V_T' = (V_T' - 2\phi_F)\left[\sqrt{1 - \frac{V_{BS}}{2\phi_F}} - 1\right] \tag{4.28}$$

Solving Eq. (4.28) for V_{BS} gives

$$V_{BS} = 2\phi_F\left[1 - \left(1 + \frac{\Delta V_T'}{V_T' - 2\phi_F}\right)^2\right]$$

Noting

$$V_{GSlat\ threshold} - V_T = V_{GSlat\ threshold}' - V_T' = \Delta V_T' = 2\ V$$

and

$$V_T' = V_T{'} - V_{FB} = -1.0 \text{ V} + 2.2 \text{ V} = 1.2 \text{ V}$$

direct substitution into the V_{BS} expression then yields

$$V_{BS} = 0.6\left[1 - \left(1 + \frac{2}{1.2 - 0.6}\right)^2\right] = -10.7 \text{ V}$$

Appendix B
Semiconductor Electrostatics —
Exact Solution

Definition of Parameters

To streamline the mathematical presentation, it is customary in the exact formulation to introduce the normalized potentials

$$U(x) = \frac{\phi(x)}{kT/q} = \frac{E_i(\text{bulk}) - E_i(x)}{kT} \tag{B.1}$$

$$U_S = \frac{\phi_s}{kT/q} = \frac{E_i(\text{bulk}) - E_i(\text{surface})}{kT} \tag{B.2}$$

and

$$U_F = \frac{\phi_F}{kT/q} = \frac{E_i(\text{bulk}) - E_F}{kT} \tag{B.3}$$

$\phi(x)$, ϕ_s, and ϕ_F were formally defined in Chapter 2 (also see Fig. 2.7). $U(x)$ is clearly the electrostatic potential normalized to kT/q and is usually referred to as "the potential" if no ambiguity exists. Similarly, $U_S = U(x = 0)$ is known as the "surface potential." U_F is simply called the doping parameter. x is of course the depth into the semiconductor as measured from the oxide–semiconductor interface. Because the electric field is assumed to vanish in the semiconductor bulk (idealization 5, Section 2.1), it is permissible to treat the semiconductor as if it extended from $x = 0$ to $x = \infty$. Note that $U(x \rightarrow \infty) = 0$ in agreement with the choice of $\phi = 0$ in the semiconductor bulk.

In addition to the normalized potentials, quantitative expressions for the band bending inside of a semiconductor are normally formulated in terms of a special length parameter known as the *intrinsic Debye length*. The Debye length is a characteristic length that was originally introduced in the study of plasmas. (A plasma is a highly ionized gas containing an equal number of positive gas ions and negative electrons.) Whenever a plasma is perturbed by placing a charge in or near it, the mobile species always rearrange so as to shield the plasma proper from the perturbing charge. The Debye length is the shielding distance, or roughly the distance where the electric field emanating from the perturbing charge falls off by a factor of $1/e$. In the bulk, or every-where under flat band conditions, the semiconductor can be viewed as a type of plasma with its equal number of ionized impurity sites and mobile electrons or holes. The placement of charge near the semiconductor, on the MOS-C gate for example, then causes the mobile species inside the semiconductor to rearrange so as to shield the semiconductor proper from the perturbing charge. The shielding distance or band-bending region is again on the order of a Debye length, the bulk or extrinsic Debye length L_B, where

$$L_B = \left[\frac{K_S \varepsilon_0 kT}{q^2 (n_{bulk} + p_{bulk})}\right]^{1/2} \tag{B.4}$$

Although the bulk Debye length characterization applies only to small deviations from flat band, it is convenient to employ the Debye length appropriate for an *intrinsic* material as a normalizing factor in theoretical expressions. The *intrinsic* Debye length, L_D, is obtained from the more general L_B relationship by setting $n_{bulk} = p_{bulk} = n_i$; that is,

$$L_D = \left[\frac{K_S \varepsilon_0 kT}{2 q^2 n_i}\right]^{1/2} \tag{B.5}$$

Exact Solution

Expressions for the charge density, electric field, and potential as a function of position inside the semiconductor are obtained by solving Poisson's equation. Since the MOS-C is assumed to be a one-dimensional structure (idealization 7, Section 2.1), Poisson's equation simplifies to

$$\frac{d\mathscr{E}}{dx} = \frac{\rho}{K_S \varepsilon_0} = \frac{q}{K_S \varepsilon_0} (p - n + N_D - N_A) \tag{B.6}$$

Maneuvering to recast the equation in a form more amenable to solution, we note

$$\mathscr{E} = \frac{1}{q} \frac{dE_i(x)}{dx} = -\frac{kT}{q} \frac{dU}{dx} \tag{B.7}$$

The first equality in Eq. (B.7) is a restatement of Eq. (3.15) in Volume I. The second equality follows from the Eq. (B.1) definition of U and the fact that $dE_i(\text{bulk})/dx = 0$. In a similar vein we can write

$$p = n_i e^{[E_i(x)-E_F]/kT} = n_i e^{U_F - U(x)} \tag{B.8a}$$

$$n = n_i e^{[E_F - E_i(x)]/kT} = n_i e^{U(x) - U_F} \tag{B.8b}$$

Moreover, since $\rho = 0$ and $U = 0$ in the semiconductor bulk,

$$0 = p_{\text{bulk}} - n_{\text{bulk}} + N_D - N_A = n_i e^{U_F} - n_i e^{-U_F} + N_D - N_A \tag{B.9}$$

or

$$N_D - N_A = n_i(e^{-U_F} - e^{U_F}) \tag{B.10}$$

Substituting the foregoing $\mathscr{E}$, p, n, and $N_D - N_A$ expressions into Eq. (B.6) yields

$$\boxed{\rho = qn_i(e^{U_F - U} - e^{U - U_F} + e^{-U_F} - e^{U_F})} \tag{B.11}$$

and

$$\frac{d^2U}{dx^2} = \left(\frac{q^2 n_i}{K_S \varepsilon_0 kT}\right)(e^{U - U_F} - e^{U_F - U} + e^{U_F} - e^{-U_F}) \tag{B.12}$$

or, in terms of the intrinsic Debye length,

$$\frac{d^2U}{dx^2} = \frac{1}{2L_D^2}(e^{U - U_F} - e^{U_F - U} + e^{U_F} - e^{-U_F}) \tag{B.13}$$

We turn next to the main task at hand. Poisson's equation, Eq. (B.13), is to be solved subject to the boundary conditions:

$$\mathscr{E} = 0 \quad \text{or} \quad \frac{dU}{dx} = 0 \quad \text{at } x = \infty \tag{B.14a}$$

and

$$U = U_S \quad \text{at } x = 0 \tag{B.14b}$$

Multiplying both sides of Eq. (B.13) by dU/dx, integrating from $x = \infty$ to an arbitrary point x, and making use of the Eq. (B.14a) boundary condition, we quickly obtain

$$\mathscr{E}^2 = \left(\frac{kT/q}{L_D}\right)^2 [e^{U_F}(e^{-U} + U - 1) + e^{-U_F}(e^U - U - 1)] \qquad (B.15)$$

Equation (B.15) is of the form $y^2 = a^2$, which has two roots, $y = a$ and $y = -a$. As can be deduced by inspection using the energy band diagram, we must have $\mathscr{E} > 0$ when $U > 0$ and $\mathscr{E} < 0$ when $U < 0$. Since the right-hand side of Eq. (B.15) is always positive ($a \geq 0$), the proper polarity for the electric field is obviously obtained by choosing the positive root when $U > 0$ and the negative root when $U < 0$. We can therefore write

$$\mathscr{E} = -\frac{kT}{q}\frac{dU}{dx} = \hat{U}_S \frac{kT}{q}\frac{F(U, U_F)}{L_D} \qquad (B.16)$$

where

$$F(U, U_F) \equiv [e^{U_F}(e^{-U} + U - 1) + e^{-U_F}(e^U - U - 1)]^{1/2} \qquad (B.17)$$

and

$$\hat{U}_S = \begin{cases} +1 & \text{if } U_S > 0 \\ -1 & \text{if } U_S < 0 \end{cases} \qquad (B.18)$$

To complete the solution, one separates the U and x variables in Eq. (B.16) and, making use of the Eq. (B.14b) boundary condition, integrates from $x = 0$ to an arbitrary point x. The end result is Eq. (B.19),

$$\hat{U}_S \int_U^{U_S} \frac{dU'}{F(U', U_F)} = \frac{x}{L_D} \qquad (B.19)$$

Although not in a totally explicit form, Eqs. (B.11), (B.16), and (B.19) collectively constitute an exact solution for the electrostatic variables. For a given U_S, numerical techniques can be used to compute U as a function of x from Eq. (B.19). Once U as a function of x is established, direct substitution into Eqs. (B.11) and (B.16) yields ρ and $\mathscr{E}$ as a function of x. The Fig. 2.8 plots of $U = \phi/(kT/q)$ versus x and ρ versus x were constructed following the cited procedure.

Appendix C
C–V Supplement

An analysis based on the exact charge distribution inside an ideal MOS-C yields the following capacitance–voltage relationships:*

$$C = \frac{C_O}{1 + \left(\dfrac{K_O W_{\text{eff}}}{K_S x_o}\right)} \tag{C.1}$$

$$W_{\text{eff}} = \begin{cases} \hat{U}_S L_D \left[\dfrac{2F(U_S, U_F)}{e^{U_F}(1 - e^{-U_S}) + e^{-U_F}(e^{U_S} - 1)}\right] & \dots \text{acc} & \text{(C.2a)} \\[3ex] \dfrac{\sqrt{2}L_D}{(e^{U_F} + e^{-U_F})^{1/2}} & \dots \text{flat band} & \text{(C.2b)} \\[3ex] \hat{U}_S L_D \left[\dfrac{2F(U_S, U_F)}{e^{U_F}(1 - e^{-U_S}) + e^{-U_F}(e^{U_S} - 1)/(1 + \Delta)}\right] & \dots \text{depl/inv} & \text{(C.2c)} \end{cases}$$

where

$$\Delta = \begin{cases} 0 & \dots \text{low frequency limit} & \text{(C.3a)} \\[3ex] \dfrac{(e^{U_S} - U_S - 1)/F(U_S, U_F)}{\displaystyle\int_{0^+}^{U_S} \dfrac{e^{U_F}(1 - e^{-U})(e^U - U - 1)}{2F^3(U, U_F)}\,dU} & \begin{array}{l}\dots \text{high frequency limit} \\ (p\text{-type MOS-C})\end{array} & \text{(C.3b)} \end{cases}$$

---

*Except for Eq. (C.3b), the relationships are valid for either n- or p-type devices. The required modification of Eq. (C.3b) for n-type devices is noted in the text. For a derivation of the low-frequency relationship see A. S. Grove, B. E. Deal, E. H. Snow, and C. T. Sah, "Investigation of Thermally Oxidised Silicon Surfaces Using Metal-Oxide-Semiconductor Structures," *Solid-State Electronics*, **8**, 145 (1965). The high-frequency result, which includes the so-called rearrangement capacitance or capacitance contribution from the movement of inversion-layer carriers, is adapted from J. R. Brews, "An Improved High-Frequency MOS Capacitance Formula," *J. Appl. Phys.*, **45**, 1276 (1974).

$$F(U, U_F) = [e^{U_F}(e^{-U} + U - 1) + e^{-U_F}(e^U - U - 1)]^{1/2} \tag{C.4}$$

$$F(U_S, U_F) = F(U = U_S, U_F) \tag{C.5}$$

$$L_D = \left[\frac{K_S \varepsilon_0 kT}{2q^2 n_i}\right]^{1/2} \tag{C.6}$$

$$U_F = \frac{\phi_F}{kT/q} \tag{C.7}$$

$$U_S = \frac{\phi_S}{kT/q} \tag{C.8}$$

$$\hat{U}_S = \begin{cases} +1 & \text{if } U_S > 0 \\ -1 & \text{if } U_S < 0 \end{cases} \tag{C.9}$$

and

$$V_G = \frac{kT}{q}\left[U_S + \hat{U}_S \frac{K_S x_o}{K_O L_D} F(U_S, U_F)\right] \tag{C.10}$$

It should be noted that $F(U, U_F)$, L_D, U_F, U_S, and $\hat{U}_S$ come from the exact solution for the semiconductor electrostatics. For additional information about the cited quantities see Appendix B.

Unlike the delta-depletion result, C cannot be expressed explicitly as a function of V_G in the exact charge formulation. Both variables, however, have been related to U_S and the capacitance expected from the structure for a given applied gate voltage can be computed numerically. The low-frequency computation is simple enough that it can be performed on a hand calculator. The usual and most efficient computational procedure is to calculate C and the corresponding V_G for a set of assumed U_S values. Typically, a sufficient set of (C, V_G) points to construct the C–V_G characteristic will be generated if U_S is stepped by whole-number units $(-5, -4, \ldots)$ over the normal operating range of U_S values $(U_F - 21 \leq U_S \leq U_F + 21$ at room temperature). It should be noted that care must be exercised if $U_S = 0$ is included as one of the computational points. At $U_S = 0$ the Eq. (C.2b) expression for W_{eff} must be employed; the accumulation and depletion/inversion relationships become indeterminate $(0/0)$ if U_S is set equal to zero. Also, the quoted high-frequency Δ-value is only valid for p-type devices. For n-type devices $[\exp(U_S) - U_S - 1] \rightarrow [\exp(-U_S) + U_S - 1]$ and $\exp(U_F)\{[1 - \exp(-U)] \cdot [\exp(U) - U - 1]\} \rightarrow \exp(-U_F)\{[\exp(U) - 1][\exp(-U) + U - 1]\}$ in Eq. (C.3b). Alternatively, it is possible to obtain an n-type characteristic by simply running the calculations for an equivalently doped p-type device and then changing the sign of all computed V_G values. The latter procedure works because of the voltage symmetry between ideal n- and p-type devices.

Appendix D
I–V Supplement

An analysis based on the exact charge distribution inside an ideal *n-channel* (*p*-bulk) MOSFET yields the following current-voltage relationship:

$$I_D\left(\begin{array}{c}\text{exact}\\\text{charge}\end{array}\right) = \frac{Z\bar{\mu}_n C_o}{L}\left[V_G(V_{SL} - V_{SO}) - \frac{1}{2}(V_{SL}^2 - V_{SO}^2)\right]$$

$$+ \frac{Z\bar{\mu}_n C_o}{L}\frac{K_S x_o}{K_O L_D}\left(\frac{kT}{q}\right)^2\left[\int_0^{U_{SO}} F(U, U_F, 0)\,dU - \int_0^{U_{SL}} F(U, U_F, U_D)\,dU\right]$$

$$(D.1)$$

where

$$F(U, U_F, \xi) \equiv [e^{U_F}(e^{-U} + U - 1) + e^{-U_F}(e^{U-\xi} - U - e^{-\xi})]^{1/2} \qquad (D.2)$$

The corresponding charge-sheet relationship is

$$I_D\left(\begin{array}{c}\text{charge}\\\text{sheet}\end{array}\right) = \frac{Z\bar{\mu}_n C_o}{L}\left\{\left(V_G + \frac{kT}{q}\right)(V_{SL} - V_{SO}) - \frac{1}{2}(V_{SL}^2 - V_{SO}^2)\right.$$

$$\left. + V_B^2\left[\sqrt{U_{SL} - 1} - \sqrt{U_{SO} - 1} - \frac{2}{3}(U_{SL} - 1)^{3/2} + \frac{2}{3}(U_{SO} - 1)^{3/2}\right]\right\}$$

$$(D.3)$$

where

$$V_B^2 \equiv \left(\frac{kT}{q}\right)^2\frac{K_S x_o}{K_O L_D}\sqrt{\frac{N_A}{n_i}} \qquad (D.4)$$

In both theories

$$\phi_F = \frac{kT}{q} U_F \qquad\qquad\qquad (D.5)$$

$$V_{S0} = \frac{kT}{q} U_{S0} \qquad\qquad\qquad (D.6)$$

$$V_{SL} = \frac{kT}{q} U_{SL} \qquad\qquad\qquad (D.7)$$

and

$$V_D = \frac{kT}{q} U_D \qquad\qquad\qquad (D.8)$$

Finally, the normalized surface potentials at the source (U_{S0}) and drain (U_{SL}) are respectively computed from

$$V_G = \frac{kT}{q} \left[U_{S0} + \frac{K_S x_o}{K_o L_D} F(U_{S0}, U_F, 0) \right] \qquad \ldots (U_{S0} > 0) \qquad (D.9a)$$

and

$$V_G = \frac{kT}{q} \left[U_{SL} + \frac{K_S x_o}{K_o L_D} F(U_{SL}, U_F, U_D) \right] \qquad \ldots (U_{SL} > 0) \qquad (D.9b)$$

To generate a set of I_D–V_D characteristics, V_G and V_D are systematically stepped over the desired range of operation. For each V_G and V_D combination, Eqs. (D.9a) and (D.9b) are iterated to determine U_{S0} and U_{SL} at the specified operating point. Once U_{S0} and U_{SL} are known, I_D can then be computed using either Eq. (D.1) or Eq. (D.3). The process is repeated of course for each V_G–V_D combination. The characteristics for a p-channel device can be established by running the calculations for an equivalently doped and biased n-channel device. Naturally, the biasing-polarities must be reversed in plotting the p-channel characteristics. For additional information about the exact-charge formalism, the reader is referred to Appendixes B and C.

The form of the relationships quoted herein are from R. F. Pierret and J. A. Shields, "Simplified Long-Channel MOSFET Theory," *Solid-State Electronics,* **26**, 143 (1983). See H. C. Pao and C. T. Sah, *Solid-State Electronics,* **9**, 927 (1966) for the original exact-charge analysis, and J. R. Brews, *Solid-State Electronics,* **21**, 345 (1978) for the original charge-sheet analysis.

Appendix E
Volume Review Problem Sets and Answers

The following problems were designed assuming a knowledge — at times an integrated knowledge — of the subject matter in Chapter 2–4. The sets could serve as a review or as a means of evaluating the reader's mastery of the subject. If the reader is well prepared, each problem set should take less than 60 minutes to complete under "closed-book" conditions. An answer key is included at the end of the problem sets. [*Note:* In answering multiple choice questions, choose the *best available* answer: select only *one* answer per question.]

PROBLEM SET A

I. MOS Fundamentals/Nonidealities

A totally dimensioned energy band diagram for an MOS-C under a specific gate bias is shown in the following figure. The device is maintained at $T = 300$ K, $kT/q = 0.0259$ V, $n_i = 1.18 \times 10^{10}/\text{cm}^3$, $K_S = 11.8$, $K_O = 3.9$, and $x_o = 0.2$ μm. Also, $Q_{IT} = 0$ and $E_F - E_i = 0$ at the surface of the semiconductor. Use the energy band diagram and the specified parametric values in answering Questions 1–12.

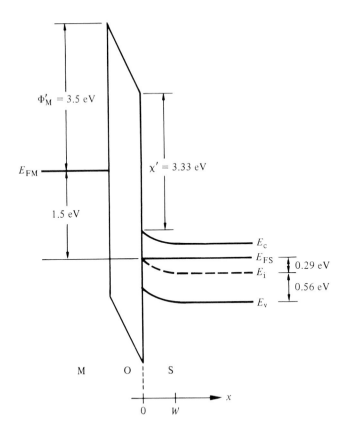

(1) The electrostatic potential (ϕ) inside the semiconductor is as sketched:

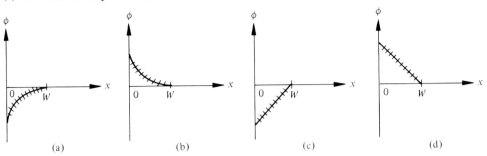

(a)   (b)   (c)   (d)

(2) The electric field ($\mathscr{E}$) inside the semiconductor is roughly as sketched next.

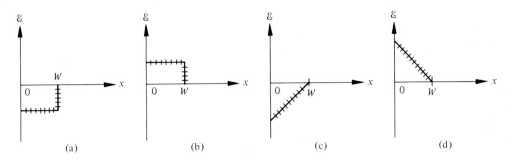

(a)   (b)   (c)   (d)

(3) Do equilibrium conditions prevail *inside the semiconductor*?

(a) Yes

(b) No

(c) Can't be determined

(4) On a semilog plot the electron concentration versus position inside the semiconductor is roughly as sketched next.

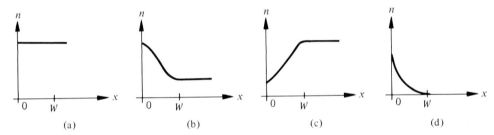

(a)   (b)   (c)   (d)

(5) N_D = ?

(a) $3.97 \times 10^{14}/cm^3$

(b) $8.60 \times 10^{14}/cm^3$

(c) $1.18 \times 10^{10}/cm^3$

(d) $1.20 \times 10^{15}/cm^3$

(6) V_G = ?

(a) 0.29 V

(b) −0.29 V

(c) 1.5 V

(d) −1.5 V

(e) 0 V

(7) What is the voltage drop ($\Delta\phi_{ox}$) across the oxide?

(a) -1.11 V

(b) -1.38 V

(c) -0.83 V

(d) -1.5 V

(e) -1.0 V

(8) What is the voltage drop across the semiconductor?

(a) -0.98 V

(b) 0.29 V

(c) -0.29 V

(d) 0.83 V

(e) -0.83 V

(9) What is the metal–semiconductor workfunction difference (ϕ_{MS})?

(a) 0.46 V

(b) 0.17 V

(c) -0.1 V

(d) -0.39 V

(10) The mobile ion charge in the oxide must be very small ($Q_M \cong 0$). One comes to this conclusion because:

(a) There is no band bending in the metal.

(b) The fields in the oxide and semiconductor have the same polarity at the oxide–semiconductor interface.

(c) The bands in the oxide are a linear function of position.

(d) The voltage drop across the oxide only slightly exceeds the voltage drop across the semiconductor.

(11) $Q_F/C_o = $? [Note: If the energy band diagram is examined very carefully, one concludes $D_{semi} = D_{ox}$ at the oxide–semiconductor interface.]

(a) Zero

(b) -0.28 V

(c) 0.58 V

(d) 0.27 V

(e) 1.5 V

(12) Compute the normalized small signal capacitance, C/C_O, at the applied bias point.

(a) 0.091

(b) 0.23

(c) 0.48

(d) 0.63

(e) 1.00

II. C–V Characteristics

(13/14) Complete the following table making use of the ideal-structure C–V characteristic and energy band diagrams included after the table. For each of the biasing conditions named in the table, employ letters (a–e) to identify the corresponding bias point on the ideal MOS-C C–V characteristic. Likewise, use a number (1–5) to identify the band diagram associated with each of the biasing conditions.

Bias condition	Capacitance (a–e)	Band diagram (1–5)
Inversion		
Depletion		
Flat band		
Threshold		
Accumulation		

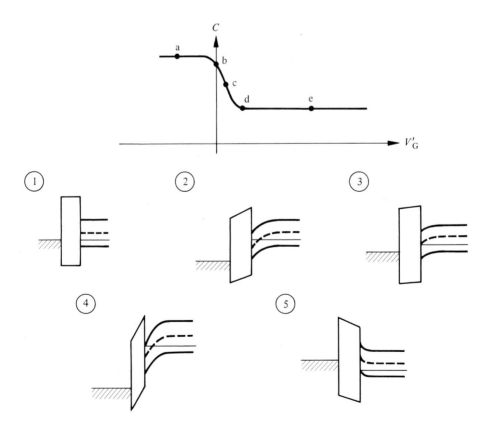

(15) Consider the C–V curves shown next. For which curve (or curves) is an *equilibrium* inversion layer present when $V_G > V_T$?

(a) Curve *a*

(b) Curve *b*

(c) Curve *c*

(d) Curves *a* and *b*

(e) Curves *b* and *c*

(f) Curves *a*, *b*, and *c*

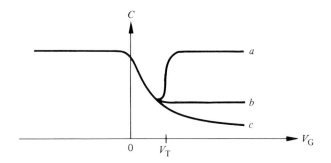

(16) Relative to the MOS-C C–V curves shown next, the MOS-C exhibiting curve *b* has:

(a) A thinner oxide

(b) A lower silicon doping

(c) Both a thinner oxide and a lower doping

(d) A thicker oxide

(e) A higher silicon doping

(f) Both a thicker oxide and a higher doping

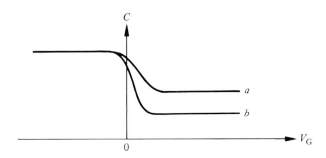

III. MOSFET

A MOSFET maintained at $T = 300$ K has an $x_o = 0.05$ μm, $N_A = 10^{16}/\text{cm}^3$, $\Phi_M = 4.1$ eV, $\chi = 4.05$ eV, $Q_F/q = 1.88 \times 10^{11}/\text{cm}^2$, and $Q_M = Q_{IT} = 0$.

(17) Calculate V_{FB}.

(a) -0.42 V

(b) -0.86 V

(c) -1.30 V

(d) 0.42 V

(18) Calculate $V_T - V_{FB}$.

(a) 0.71 V

(b) 1.41 V

(c) 0.35 V

(d) 1.06 V

(19) Calculate the depletion width at the source end of the channel ($y = 0$) when $V_G = V_T$.

(a) 0.304 μm

(b) 0.215 μm

(c) 0.608 μm

(d) 1.22 μm

(20) Based on the square-law theory, which of the following plots correctly characterizes g_m when $V_D > V_{Dsat}$?

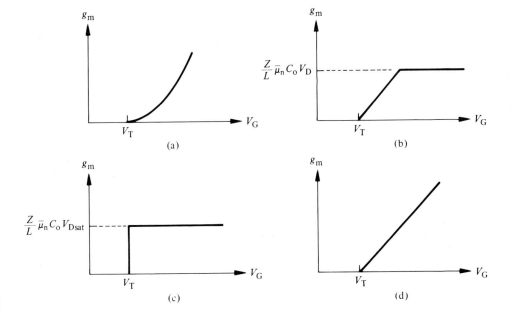

IV. True or False

(21) The effective mobility of carriers in MOSFET surface channels is always less than or equal to the bulk mobility of the same carriers.

(a) True

(b) False

(22) The fixed oxide charge in MOS devices is typically distributed almost uniformly throughout the oxide.

(a) True

(b) False

(23) The fixed oxide charge is typically minimized by annealing the structure in the presence of hydrogen at temperatures less than 500° C.

(a) True

(b) False

(24) "Saturation" in a MOSFET occurs when both the source and drain inject minority carriers into the channel region at the same time.

(a) True

(b) False

(25) The "square-law" theory for the dc characteristics of a MOSFET derives its name from the fact that, in this theory, one obtains $I_{Dsat} \propto (V_G - V_T)^2$.

(a) True

(b) False

(26) MOSFETs that do not conduct current (are "off") when $V_G = 0$ are typically referred to as "depletion-mode" devices.

(a) True

(b) False

(27) "Field-effect" is the term used to describe the elimination of the conducting inversion layer at the drain end of the MOSFET channel when $V_D = V_{Dsat}$.

(a) True

(b) False

(28) Both the fixed-charge and the interfacial-trap density are greater on {111} Si surfaces than on {100} Si surfaces.

(a) True

(b) False

(29) "Bias-temperature stressing" in MOS work refers to applying a bias between the source and drain of a MOSFET while monitoring the temperature rise of the substrate.

(a) True

(b) False

(30) Ion implantation and back biasing are two methods that have been employed to adjust the threshold voltage of MOSFETs.

(a) True

(b) False

PROBLEM SET B

I. MOS Fundamentals

A totally dimensioned energy band diagram for an M"O"S-C recently fabricated in a research laboratory is shown below. (The "O" is actually ZnSe and the semiconductor is GaAs.) The device is maintained at $T = 300$ K, $kT/q = 0.0259$ V, $n_i = 2.25 \times 10^6/\text{cm}^3$, $K_S = 12.85$, $K_O = 9.0$, and $x_o = 0.1$ μm. It has also been established that $Q_M = 0$, $Q_F = 0$, and $Q_{IT} = 0$. Use the cited energy band diagram and the given information in answering Questions 1–10.

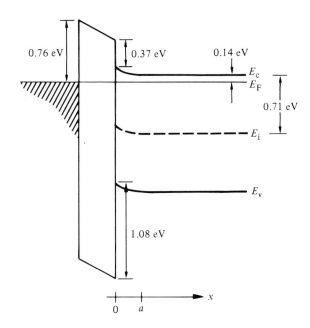

(1) Sketch the electrostatic potential (ϕ) inside the semiconductor as a function of position. (Let $\phi = 0$ in the semiconductor bulk.)

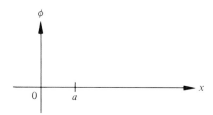

(2) Roughly sketch the electric field ($\mathcal{E}$) inside the semiconductor as a function of position.

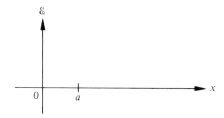

(3) Do equilibrium conditions prevail *inside the semiconductor*?

(a) Yes

(b) No

(c) Can't be determined

(4) $N_D = ?$

(a) $4.03 \times 10^{20}/\text{cm}^3$

(b) $8.13 \times 10^{15}/\text{cm}^3$

(c) $1.00 \times 10^{15}/\text{cm}^3$

(d) $5.01 \times 10^8/\text{cm}^3$

(5) $V_G = ?$

(a) -0.57 V

(b) -0.39 V

(c) 0 V

(d) 0.39 V

(e) 0.57 V

(6) For the pictured condition the M"O"S-C is

(a) Accumulated.

(b) Depleted.

(c) Inverted.

(d) Biased at the depletion–inversion transition point.

(7) What is the metal–semiconductor workfunction difference (ϕ_{MS})?

(a) −0.39 V

(b) −0.25 V

(c) 0 V

(d) 0.25 V

(e) 0.39 V

(8) What voltage must be applied to the gate to achieve flat band conditions?

(a) −0.39 V

(b) −0.25 V

(c) 0 V

(d) 0.25 V

(e) 0.39 V

(9) Invoking the delta-depletion approximation, determine the normalized small signal capacitance, C/C_O, at the pictured bias point.

(a) 0.25

(b) 0.41

(c) 0.56

(d) 0.83

(10) As noted in the energy band figure, a is the distance from the "oxide"–semiconductor interface to the quasi-neutral semiconductor bulk. Determine the length of a at the pictured bias point.

(a) 0.112 μm

(b) 0.205 μm

(c) 0.428 μm

(d) 0.813 μm

II. MOSFET

A standard MOSFET is fabricated with $\phi_{MS} = -0.89$ V, $Q_M = 0$, $Q_{IT} = 0$, $Q_F/q = 5 \times 10^{10}/cm^2$, $x_o = 500$ Å, $A_G = 10^{-3}$ cm², and $N_A = 10^{15}/cm^3$. Assume $T = 300$ K.

(11) Determine the flat band gate voltage, V_{FB}.

(a) -2.05 V

(b) -1.01 V

(c) -0.89 V

(d) 0 V

(12) Determine the gate voltage at the onset of inversion, V_T.

(a) -1.01 V

(b) -0.22 V

(c) 0.79 V

(d) 1.80 V

(13) The given MOSFET is

(a) An enhancement-mode MOSFET.

(b) A depletion-mode MOSFET.

(c) A built-in channel MOSFET.

(14) If the internal condition inside the MOSFET is as shown below to the left, identify the corresponding operational point on the I_D–V_D characteristic at the right.

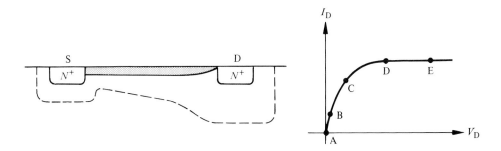

(15) At $V_G - V_T = 3$ V and $V_D = 1$ V the MOSFET exhibits a drain current of $I_D = 2.5 \times 10^{-4}$ amp. Using the square-law formulation, determine the drain current if $V_G - V_T = 3$ V and $V_D = 4$ V.

(a) 3.5×10^{-4} amp

(b) 4.0×10^{-4} amp

(c) 4.5×10^{-4} amp

(d) 1.0×10^{-3} amp

III. True or False

(16) The "field effect" is the phenomenon where carriers are accelerated by an electric field impressed parallel to the surface of the semiconductor.

(a) True

(b) False

(17) The electron affinity (χ) of a semiconductor is the difference in energy between the vacuum level and E_c at the surface of the semiconductor.

(a) True

(b) False

(18) The "quasi-static technique" is employed in measuring the low-frequency MOS-C $C-V$ characteristics.

(a) True

(b) False

(19) The nonequilibrium condition where there is a deficit of minority carriers and a depletion width in excess of the equilibrium value is referred to as "deep inversion."

(a) True

(b) False

(20) The voltage shift due to mobile ions in the oxide is at a minimum when the ions are located midway between the gate and semiconductor.

(a) True

(b) False

(21) The interfacial trap charge (Q_{IT}) is typically a function of the applied gate voltage.

(a) True

(b) False

(22) The "bulk-charge" theory for the dc characteristics of a MOSFET derives its name from the fact that, in this theory, one properly accounts for changes in the "bulk" or depletion-region charge beneath the MOSFET channel.

(a) True

(b) False

(23) Let g_d be the drain or channel conductance of a MOSFET. By definition, at low frequencies $g_d = \partial I_D / \partial V_G |_{V_D = \text{constant}}$.

(a) True

(b) False

(24) The mobility of carriers in surface inversion layers or channels is typically lower than the bulk mobility of the same carriers because of the added scattering associated with the depletion region charge.

(a) True

(b) False

(25) In modern-day MOS structures the "M" in MOS is often heavily doped polycrystalline Si.

(a) True

(b) False

ANSWERS — SET A

(1) a . . . ϕ has the same shape as the "upsidedown" of the bands.

(2) c . . . $\mathscr{E}$ is proportional to the slope of the bands.

(3) a . . . E_F is energy-independent inside the semiconductor.

(4) c . . . The semiconductor is depleted, has a lower n, near the surface. $n \cong N_D$ for $x > W$.

(5) b . . . $N_D = n_i e^{(E_F - E_i)/kT} = (1.18 \times 10^{10}) e^{0.29/0.0259} = 8.60 \times 10^{14}/\text{cm}^3$.

(6) d . . . Per Eq. (2.1), $V_G = -(1/q)[E_F(\text{metal}) - E_F(\text{semi})] = -1.5$ V.

(7) a . . . Using the semiconductor Fermi level as a reference,

$$\Delta\phi_{OX} = -\frac{1}{q}[\Phi'_M + (E_{FM} - E_{FS}) - \chi' - E_G/2]$$

$$= 3.5 + 1.5 - 3.33 - 0.56 = -1.11 \text{ V}$$

(8) c . . . Per Eq. (2.5), $\phi_S = (1/q)[E_i(\text{bulk}) - E_i(\text{surface})] = -0.29$ V.

(9) c . . . $\phi_{MS} = (1/q)[\Phi'_M - \chi' - (E_c - E_F)_\infty] = 3.5 - 3.33 - (0.56 - 0.29) = -0.10$ V

(10) c

(11) a . . . If $D_{semi} = D_{OX}$, there is no charge at the interface and $Q_F = 0$.

(12) c . . . $W = \left[\dfrac{2K_S\varepsilon_0}{qN_D}\phi_S\right]^{1/2} = \left[\dfrac{(2)(11.8)(8.85 \times 10^{-14})(0.29)}{(1.6 \times 10^{-19})(8.60 \times 10^{14})}\right]^{1/2} = 0.664 \ \mu\text{m}$

$$\frac{C}{C_0} = \frac{1}{1 + \left(\dfrac{K_O W}{K_S x_o}\right)} = \frac{1}{1 + \left[\dfrac{(3.9)(0.664)}{(11.8)(0.2)}\right]} = 0.477$$

(13/14) Inversion   . . . e, 4

Depletion   . . . c, 3

Flat band   . . . b, 1

Threshold   . . . d, 2

Accumulation   . . . a, 5

(15) d . . . Curve-c is a nonequilibrium deep-depletion characteristic.

(16) b . . . In accumulation $C \rightarrow C_O = K_O \varepsilon_0 A_G / x_o$. Since both devices exhibit the same capacitance in accumulation, the two devices have the same oxide thickness. The lower capacitance of device-b in inversion indicates this device has a lower doping. (W_T decreases with increasing doping, thereby giving rise to a larger capacitance; also see Fig. 2.14.)

(17) c . . . $V_{FB} = \phi_{MS} - Q_F / C_o$

$$\phi_{MS} = (1/q)[\Phi'_M - \chi' - (E_c - E_F)_x] = (1/q)[\Phi_M - \chi - (E_c - E_F)_x]$$

$$(E_c - E_F)_x = (E_c - E_i) + (E_i - E_F)_x$$

$$\cong \frac{E_G}{2} + kT \ln\left(\frac{N_A}{n_i}\right) = 0.56 + 0.0259 \ln\left(\frac{10^{16}}{1.18 \times 10^{10}}\right)$$

$$= 0.91 \text{ eV}$$

$$\phi_{MS} = 4.1 - 4.05 - 0.91 = -0.86 \text{ V}$$

$$-\frac{Q_F}{C_o} = -q\frac{x_o}{K_O \varepsilon_0}\frac{Q_F}{q} = -\frac{(1.6 \times 10^{-19})(5 \times 10^{-6})(1.88 \times 10^{11})}{(3.9)(8.85 \times 10^{-14})}$$

$$= -0.44 \text{ V}$$

$$V_{FB} = -0.86 - 0.44 = -1.30 \text{ V}$$

(18) b . . . $V_T - V_{FB} = V'_T = 2\phi_F + \frac{K_S}{K_O}x_o\sqrt{\frac{4qN_A}{K_S \varepsilon_0}\phi_F}$

$$\phi_F = \frac{kT}{q}\ln\left(\frac{N_A}{n_i}\right) = 0.0259 \ln\left(\frac{10^{16}}{1.18 \times 10^{10}}\right) = 0.354 \text{ V}$$

$$V_T - V_{FB} = 2(0.354) + \frac{(11.8)(5 \times 10^{-6})}{(3.9)}\left[\frac{(4)(1.6 \times 10^{-19})(10^{16})(0.354)}{(11.8)(8.85 \times 10^{-14})}\right]^{1/2}$$

$$= 1.41 \text{ V}$$

(19) a . . . With the source grounded, the source side of the channel is always at equilibrium. Since we are given $V_G = V_T$, it follows that

$$W = \left[\frac{2K_S \varepsilon_0}{qN_A}(2\phi_F)\right]^{1/2} = \left[\frac{(2)(11.8)(8.85 \times 10^{-14})(2)(0.354)}{(1.6 \times 10^{-19})(10^{16})}\right]^{1/2}$$

$$= 0.304 \ \mu\text{m}$$

(20) d . . . Given $V_D > V_{Dsat}$ we know the device is saturation biased. In the square-law formulation

$$I_D = I_{Dsat} = \frac{Z\bar{\mu}_n C_o}{2L}(V_G - V_T)^2$$

$$g_m = \left.\frac{\partial I_D}{\partial V_G}\right|_{V_D} = \frac{Z\bar{\mu}_n C_o}{2L}(V_G - V_T)$$

For $V_G < V_T$ the device is off and $g_m = 0$. g_m is seen to be a linear function of V_G for $V_G > V_T$. (Actually, $\bar{\mu}_n$ will decrease slightly with increasing V_G giving rise to a slope-over in the characteristic. In any case, answer d is closest to the expected shape.)

(21) a

(22) b

(23) b

(24) b

(25) a

(26) b

(27) b

(28) a

(29) b

(30) a

ANSWERS — SET B

(1) . . . ϕ has the same shape as the "upsidedown" of the bands.

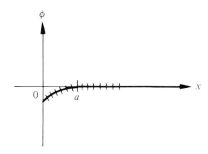

(2) ...$\mathcal{E}$ is proportional to the slope of the bands.

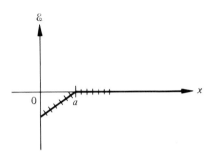

(3) a...The device is zero biased, thus the semiconductor must be in equilibrium. E_F is indeed shown energy-independent inside the semiconductor.

(4) b...$N_D = n_i e^{(E_F - E_i)/kT} = (2.25 \times 10^6) e^{0.57/0.0259} = 8.13 \times 10^{15}/\text{cm}^3$.

(5) c...Per Eq. (2.1), $V_G = -(1/q)[E_F(\text{metal}) - E_F(\text{semi})] = 0$.

(6) b

(7) d...$\phi_{MS} = (1/q)[\Phi'_M - \chi' - (E_c - E_F)_\infty] = 0.76 - 0.37 - 0.14 = 0.25$ V

(8) d...Since $Q_M = Q_F = Q_{IT} = 0$, $\Delta V_G = (V_G - V'_G)|_{\text{same } \phi_S} = \phi_{MS}$.
Under flat band conditions $V'_G = 0$ and $V_G = V_{FB} = \phi_{MS} = 0.25$ V.

(9) c...The delta-depletion analysis gave [Eq. (2.37)],

$$\frac{C}{C_O} = \frac{1}{\sqrt{1 + V'_G/V_\delta}}$$

It should be pointed out that V_G, not V'_G, was actually used in text Eq. (2.37). However, an ideal structure was assumed throughout Chapter 2 (all the V_G in Chapter 2 are in reality V'_G). As noted in the answer to Question 8, $\Delta V_G = (V_G - V'_G)|_{\text{same } \phi_S} = \phi_{MS}$. Since $V_G = 0$ for the pictured bias point, the corresponding $V'_G = -\phi_{MS}$.

$$V_\delta = -\frac{q}{2}\frac{K_S x_0^2}{K_O^2 \varepsilon_0} N_D = -\frac{(1.6 \times 10^{-19})(12.85)(10^{-5})^2(8.13 \times 10^{15})}{(2)(9)^2(8.85 \times 10^{-14})} = -0.117 \text{ V}$$

$$\frac{C}{C_O} = \frac{1}{\sqrt{1 + \dfrac{0.25}{0.117}}} = 0.56$$

(10) a...The quantity a is of course just the depletion width W. Since C/C_O was determined in Question 9, W can be computed using Eq. (2.34b).

$$\frac{C}{C_O} = \frac{1}{1 + \left(\dfrac{K_O W}{K_S x_o}\right)}$$

$$W = \frac{K_S}{K_O} x_o \left(\frac{C_O}{C} - 1\right) = \frac{(12.85)\,(10^{-5})}{(9)} \left(\frac{1}{0.56} - 1\right) = 0.112 \ \mu\text{m}$$

(11) b ... $V_{FB} = \phi_{MS} - \dfrac{Q_F}{C_o} = \phi_{MS} - q\dfrac{x_o}{K_O \varepsilon_0}\dfrac{Q_F}{q}$

$$= -0.89 - \frac{(1.6 \times 10^{-19})\,(5 \times 10^{-6})\,(5 \times 10^{10})}{(3.9)\,(8.85 \times 10^{-14})}$$

$$= -1.01 \ \text{V}$$

(12) b ... $V_T = V'_T + V_{FB}$

$$V'_T = 2\phi_F + \frac{K_S}{K_O} x_o \sqrt{\frac{4qN_A}{K_S \varepsilon_0} \phi_F}$$

$$\phi_F = \frac{kT}{q} \ln\left(\frac{N_A}{n_i}\right) = 0.0259 \ \ln\left(\frac{10^{15}}{1.18 \times 10^{10}}\right) = 0.294 \ \text{V}$$

$$V'_T = 2(0.294) + \frac{(11.8)\,(5 \times 10^{-6})}{(3.9)} \left[\frac{(4)\,(1.6 \times 10^{-19})\,(10^{15})\,(0.294)}{(11.8)\,(8.85 \times 10^{-14})}\right]^{1/2}$$

$$= 0.791 \ \text{V}$$

$$V_T = 0.79 - 1.01 = -0.22 \ \text{V}$$

(13) b ... Since the MOSFET conducts for $V_G > V_T$, the device is "on" at $V_G = 0$ and is therefore a depletion-mode MOSFET.

(14) D ... The MOSFET channel is shown just being pinched-off. This corresponds to the start of saturation, point D.

(15) c ... If $V_G - V_T = 3$ V and $V_D = 1$ V, the MOSFET is biased below pinch-off and, in the square-law formulation,

$$I_D = \frac{Z\bar{\mu}_n C_o}{L}\left[(V_G - V_T)V_D - \frac{V_D^2}{2}\right]$$

or

$$\frac{Z\bar{\mu}_n C_o}{L} = \frac{I_D}{(V_G - V_T)V_D - V_D^2/2} = \frac{2.5 \times 10^{-4}}{3 - 0.5} = 10^{-4} \ \text{amps/volt}^2$$

When $V_G - V_T = 3$ V and $V_D = 4$ V, the device is saturation biased, and

$$I_D = I_{Dsat} = \frac{Z\bar{\mu}_n C_o}{2L}(V_G - V_T)^2 = \frac{(10^{-4})(3)^2}{2}$$

$$= 4.5 \times 10^{-4} \text{ amps}$$

(16) b

(17) a

(18) a

(19) b

(20) b

(21) a

(22) a

(23) b

(24) b

(25) a

Appendix F
List of Symbols

a	half-width of the channel region in a J-FET
A_G	gate area
C	capacitance; MOS-C capacitance
C_G	MOSFET gate capacitance
C_{gd}	gate-to-drain capacitance in the high-frequency, small-signal equivalent circuit for the MOSFET
C_{gs}	gate-to-source capacitance in the high-frequency, small-signal equivalent circuit for the MOSFET
C_o	oxide capacitance per unit area (pf/cm^2)
C_O	$C_O = C_o A_G$; oxide capacitance (pf)
C_S	semiconductor capacitance
D	drain
D_{IT}	density of interfacial traps (states/cm^2-eV)
D_N	electron diffusion coefficient
D_{ox}	dielectric displacement in the oxide
D_{semi}	dielectric displacement in the semiconductor
$\mathcal{E}, \mathscr{E}$	electric field
$\mathcal{E}_{ox}$	electric field in the oxide
$\mathcal{E}_S$	surface electric field; electric field in the semiconductor at the oxide–semiconductor interface
$\mathcal{E}_{vac}$	electric field in a vacuum
$\mathcal{E}_y$	y-direction component of the electric field
E_c	minimum conduction band energy
E_F	Fermi energy or Fermi level
E_i	intrinsic Fermi level
E_v	maximum valence band energy

E_{vacuum}	vacuum level, minimum energy an electron must possess to completely free itself from a material
f	frequency (Hz)
f_{max}	maximum operational frequency of a MOSFET, cutoff frequency
$F(U, U_F)$	field function [see Eq. (B.17)]
G	gate
G_0	channel conductance one would observe in a J-FET if there were no depletion regions
g_d	drain or channel conductance
g_m	transconductance or mutual conductance
i_d	small signal drain current
I_D	dc drain current in a J-FET or MOSFET
I_{D0}	$V_G = 0$ saturation drain current in a J-FET
I_{Dsat}	saturation drain current
j	$\sqrt{-1}$
$\mathbf{J}_N, J_N$	electron current density
J_{Ny}	y-direction component of the electron current denisty
k	Boltzmann constant (8.62×10^{-5} eV/K)
K_O	oxide dielectric constant
K_S	semiconductor dielectric constant
L	length of the J-FET or MOSFET channel
L'	reduced channel length defined in Fig. 5.4
L_B	extrinsic Debye length
L_D	intrinsic Debye length
L_{min}	minimum MOSFET channel length yielding long-channel behavior
n	electron carrier concentration (number of electrons/cm^3)
N_A	total number of acceptor atoms or sites/cm^3
N_B	bulk semiconductor doping (N_A or N_D as appropriate)
n_{bulk}	electron concentration in the semiconductor bulk
N_D	total number of donor atoms or sites/cm^3
n_i	intrinsic carrier concentration
N_I	number of implanted ions/cm^2
p	hole concentration (number of holes/cm^3)
p_{bulk}	hole concentration in the semiconductor bulk
p_s	hole concentration at the semiconductor surface (number/cm^3)
q	magnitude of the electronic charge (1.60×10^{-19} coul)
Q	general designation for a charge
Q_B	bulk or depletion-region charge per unit area of the gate
Q_{BL}	Q_B in a long-channel MOSFET
Q_{BS}	Q_B in a short-channel MOSFET

Q_F	fixed-oxide charge per unit area at the oxide–semiconductor interface
Q_I	implant-related charge/cm^2 located at the oxide–semiconductor interface
Q_{IT}	net charge per unit area associated with the interfacial traps
Q_M	total mobile ion charge within the oxide per unit area of the gate
Q_N	total electronic charge/cm^2 in the MOSFET channel
$Q_{0\text{-}S}$	charge per unit area located at the oxide–semiconductor interface
Q_S	total charge in the semiconductor per unit area of the gate
R	ramp rate (see Fig. 2.17)
R_D	channel-to-drain resistance in a J-FET
r_j	depth of the source and drain islands in a MOSFET
R_S	source-to-channel resistance in a J-FET
S	source
T	temperature
U	electrostatic potential normalized to kT/q
U_F	semiconductor doping parameter
U_S	normalized surface potential, U evaluated at the oxide–semiconductor interface
$\hat{U}_S$	sign ($\pm$) of U_S
V	voltage, electrostatic potential
V_A	voltage applied across a pn junction
V_{bi}	"built-in" pn junction voltage
V_{BS}	back-to-source voltage
v_d	ac drain voltage
V_D	dc drain voltage
V_{DS}	drain-to-source voltage
v_{dsat}	saturation drift velocity
V_{Dsat}	saturation drain voltage
V_{FB}	flat band voltage
v_g	ac gate voltage
V_G	dc gate voltage
V'_G	dc gate voltage applied to an ideal device
V'_{GB}	gate-to-back voltage being applied to an ideal device
V_{GS}	gate-to-source voltage
V'_{GS}	gate-to-source voltage being applied to an ideal device
V_P	pinch-off gate voltage in a J-FET
V_T	inversion-depletion transition point gate voltage; MOSFET threshold or turn-on voltage
V'_T	ideal device inversion-depletion transition point gate voltage
V_W	defined voltage [see Eq. (3.24)]
V_δ	defined voltage [see Eq. (2.36)]
W	depletion width

W_D	drain *pn* junction depletion width in a MOSFET
W_{eff}	effective depletion width
W_S	source *pn* junction depletion width in a MOSFET
W_T	depletion width when the semiconductor is biased at the inversion–depletion transition point
x_c	depth of the MOSFET channel
x_o	oxide thickness
Z	width of the J-FET or MOSFET channel
Δ	frequency parameter in the exact-charge C–V theory [see Eqs. (C.3)]
ΔL	decrease in the channel length under post pinch-off conditions
ΔQ	general designation for a change in charge
ΔQ_{gate}	change in the gate charge/cm^2
ΔQ_{semi}	change in the charge/cm^2 inside the semiconductor
ΔV_G	difference between the actual device and ideal device gate voltage required to achieve a given semiconductor surface potential
ΔV_T	change in the threshold voltage due to small-dimension effects
$\Delta V_T'$	change in the threshold voltage due to back biasing (specifically applied to ideal devices)
$\Delta \phi_{ox}$	voltage drop across the oxide
$\Delta \phi_{semi}$	voltage drop across the semiconductor
γ_M	normalized centroid of mobile ion charge in the oxide
ε	permittivity
ε_0	permittivity of free space (8.85×10^{-14} farad/cm)
μ_{bulk}	carrier mobility in the bulk of a semiconductor
μ_n	electron mobility
$\bar{\mu}_n$	effective electron mobility
$\bar{\mu}_p$	effective hole mobility
ρ	charge density (coul/cm^3)
ρ_{ion}	ionic charge density inside the oxide
ρ_{ox}	charge density in the oxide
ϕ	electrostatic potential inside the semiconductor component of an MOS device structure
ϕ_F	reference voltage related to the semiconductor doping concentration
Φ_M	metal workfunction
Φ_M'	$\Phi_M' = \Phi_M - \chi_i$; effective metal workfunction in an MOS structure
ϕ_{MS}	metal–semiconductor workfunction difference
ϕ_{ox}	voltage inside the oxide
ϕ_S	semiconductor surface potential
Φ_S	semiconductor workfunction
χ	semiconductor electron affinity

χ'	$\chi' = \chi - \chi_i$; effective semiconductor electron affinity in an MOS structure
χ_i	insulator (oxide) electron affinity
χ_{Si}	silicon electron affinity
ω	radian frequency

Index